2016 SQA Past Papers & Hodder Gibson Model Papers With Answers

Advanced Higher
BIOLOGY

2015 Specimen Question Paper, Model Paper & 2016 Exam

HODDER GIBSON
AN HACHETTE UK COMPANY

This book contains the official 2015 SQA Specimen Question Paper and the 2016 Exam for Advanced Higher Biology, with associated SQA-approved answers modified from the official marking instructions that accompany the paper.

In addition the book contains a model paper, together with answers, plus study skills advice. This paper, which may include a limited number of previously published SQA questions, has been specially commissioned by Hodder Gibson, and has been written by experienced senior teachers and examiners in line with the new Advanced Higher for CfE syllabus and assessment outlines. This is not SQA material but has been devised to provide further practice for Advanced Higher examinations in 2016 and beyond.

Hodder Gibson is grateful to the copyright holders, as credited on the final page of the Answer Section, for permission to use their material. Every effort has been made to trace the copyright holders and to obtain their permission for the use of copyright material. Hodder Gibson will be happy to receive information allowing us to rectify any error or omission in future editions.

Hachette UK's policy is to use papers that are natural, renewable and recyclable products and made from wood grown in sustainable forests. The logging and manufacturing processes are expected to conform to the environmental regulations of the country of origin.

Orders: please contact Bookpoint Ltd, 130 Park Drive, Milton Park, Abingdon, Oxon OX14 4SE. Telephone: (44) 01235 827720. Fax: (44) 01235 400454. Lines are open 9.00–5.00, Monday to Saturday, with a 24-hour message answering service. Visit our website at www.hoddereducation.co.uk. Hodder Gibson can be contacted direct on: Tel: 0141 333 4650; Fax: 0141 404 8188; email: hoddergibson@hodder.co.uk

This collection first published in 2016 by
Hodder Gibson, an imprint of Hodder Education,
An Hachette UK Company
211 St Vincent Street
Glasgow G2 5QY

Typeset by Aptara, Inc.

Printed in the UK

A catalogue record for this title is available from the British Library

ISBN: 978-1-4718-9075-8

3 2 1

2017 2016

Introduction

Study Skills – what you need to know to pass exams!

Pause for thought

Many students might skip quickly through a page like this. After all, we all know how to revise. Do you really though?

Think about this:

"IF YOU ALWAYS DO WHAT YOU ALWAYS DO, YOU WILL ALWAYS GET WHAT YOU HAVE ALWAYS GOT."

Do you like the grades you get? Do you want to do better? If you get full marks in your assessment, then that's great! Change nothing! This section is just to help you get that little bit better than you already are.

There are two main parts to the advice on offer here. The first part highlights fairly obvious things but which are also very important. The second part makes suggestions about revision that you might not have thought about but which WILL help you.

Part 1

DOH! It's so obvious but …

Start revising in good time

Don't leave it until the last minute – this will make you panic.

Make a revision timetable that sets out work time AND play time.

Sleep and eat!

Obvious really, and very helpful. Avoid arguments or stressful things too – even games that wind you up. You need to be fit, awake and focused!

Know your place!

Make sure you know exactly **WHEN and WHERE** your exams are.

Know your enemy!

Make sure you know what to expect in the exam.

How is the paper structured?

How much time is there for each question?

What types of question are involved?

Which topics seem to come up time and time again?

Which topics are your strongest and which are your weakest?

Are all topics compulsory or are there choices?

Learn by DOING!

There is no substitute for past papers and practice papers – they are simply essential! Tackling this collection of papers and answers is exactly the right thing to be doing as your exams approach.

Part 2

People learn in different ways. Some like low light, some bright. Some like early morning, some like evening or night. Some prefer warm, some prefer cold. But everyone uses their BRAIN and the brain works when it is active. Passive learning – sitting gazing at notes – is the most INEFFICIENT way to learn anything. Below you will find tips and ideas for making your revision more effective and maybe even more enjoyable. What follows gets your brain active, and active learning works!

Activity 1 – Stop and review

Step 1

When you have done no more than 5 minutes of revision reading STOP!

Step 2

Write a heading in your own words which sums up the topic you have been revising.

Step 3

Write a summary of what you have revised in no more than two sentences. Don't fool yourself by saying, "I know it, but I cannot put it into words". That just means you don't know it well enough. If you cannot write your summary, revise that section again, knowing that you must write a summary at the end of it. Many of you will have notebooks full of blue/black ink writing. Many of the pages will not be especially attractive or memorable so try to liven them up a bit with colour as you are reviewing and rewriting. **This is a great memory aid, and memory is the most important thing.**

Activity 2 – Use technology!

Why should everything be written down? Have you thought about "mental" maps, diagrams, cartoons and colour to help you learn? And rather than write down notes, why not record your revision material?

What about having a text message revision session with friends? Keep in touch with them to find out how and what they are revising and share ideas and questions.

Why not make a video diary where you tell the camera what you are doing, what you think you have learned and what you still have to do? No one has to see or hear it, but the process of having to organise your thoughts in a formal way to explain something is a very important learning practice.

Be sure to make use of electronic files. You could begin to summarise your class notes. Your typing might be slow, but it will get faster and the typed notes will be easier to read than the scribbles in your class notes. Try to add different fonts and colours to make your work stand out. You can easily Google relevant pictures, cartoons and diagrams which you can copy and paste to make your work more attractive and **MEMORABLE**.

Activity 3 – This is it. Do this and you will know lots!

Step 1

In this task you must be very honest with yourself! Find the SQA syllabus for your subject (www.sqa.org.uk). Look at how it is broken down into main topics called MANDATORY knowledge. That means stuff you MUST know.

Step 2

BEFORE you do ANY revision on this topic, write a list of everything that you already know about the subject. It might be quite a long list but you only need to write it once. It shows you all the information that is already in your long-term memory so you know what parts you do not need to revise!

Step 3

Pick a chapter or section from your book or revision notes. Choose a fairly large section or a whole chapter to get the most out of this activity.

With a buddy, use Skype, Facetime, Twitter or any other communication you have, to play the game "If this is the answer, what is the question?". For example, if you are revising Geography and the answer you provide is "meander", your buddy would have to make up a question like "What is the word that describes a feature of a river where it flows slowly and bends often from side to side?".

Make up 10 "answers" based on the content of the chapter or section you are using. Give this to your buddy to solve while you solve theirs.

Step 4

Construct a wordsearch of at least 10 × 10 squares. You can make it as big as you like but keep it realistic. Work together with a group of friends. Many apps allow you to make wordsearch puzzles online. The words and phrases can go in any direction and phrases can be split. Your puzzle must only contain facts linked to the topic you are revising. Your task is to find 10 bits of information to hide in your puzzle, but you must not repeat information that you used in Step 3. DO NOT show where the words are. Fill up empty squares with random letters. Remember to keep a note of where your answers are hidden but do not show your friends. When you have a complete puzzle, exchange it with a friend to solve each other's puzzle.

Step 5

Now make up 10 questions (not "answers" this time) based on the same chapter used in the previous two tasks. Again, you must find NEW information that you have not yet used. Now it's getting hard to find that new information! Again, give your questions to a friend to answer.

Step 6

As you have been doing the puzzles, your brain has been actively searching for new information. Now write a NEW LIST that contains only the new information you have discovered when doing the puzzles. Your new list is the one to look at repeatedly for short bursts over the next few days. Try to remember more and more of it without looking at it. After a few days, you should be able to add words from your second list to your first list as you increase the information in your long-term memory.

FINALLY! Be inspired...

Make a list of different revision ideas and beside each one write **THINGS I HAVE** tried, **THINGS I WILL** try and **THINGS I MIGHT** try. Don't be scared of trying something new.

And remember – "FAIL TO PREPARE AND PREPARE TO FAIL!"

Advanced Higher Biology

The practice papers in this book give an overall and comprehensive coverage of assessment of **Knowledge** and skills of **Scientific Inquiry** for the new CfE Advanced Higher Biology.

We recommend that you download and print a copy of the Advanced Higher Biology Course Assessment Specification (CAS) pages 8–17 from the SQA website at www.sqa.org.uk.

The Course

The Advanced Higher Biology Course consists of three National Units. These are Cells and Proteins, Organisms and Evolution, and Investigative Biology. In each of the Units you will be assessed on your ability to demonstrate and apply knowledge of Biology and to demonstrate and apply skills of scientific inquiry.

You must also complete a project, the purpose of which is to allow you to carry out an in-depth investigation of a Biology topic and produce a project–report. You will also take a Course examination.

How the Course is graded

To achieve a Course award for Advanced Higher Biology you must pass all three National Unit Assessments which will be assessed by your school or college on a pass or fail basis. The grade you get depends on the following two Course assessments, which are set and graded by SQA.

1. The project is worth 25% of the grade and is marked out of 30 marks. The majority of the marks will be awarded for applying scientific inquiry skills. The other marks will be awarded for applying related knowledge and understanding.

2. A written Course examination is worth the remaining 75% of the grade. The examination is marked out of 90 marks, 60–70 of which are for the demonstration and application of knowledge with the balance for skills of scientific inquiry.

This book should help you practise the examination part! To pass Advanced Higher Biology with a C grade you will need about 50% of the 120 marks available for the project and the Course examination combined. For a B you will need roughly 60% and, for an A, roughly 70% of the marks available.

The Course examination

The Course examination is a single question paper divided into two sections.

- The first section is an objective test with 25 multiple choice items worth 25 marks.

- The second section is a mix of restricted and extended response questions worth between 1 and 9 marks each for a total of 65 marks. The majority of the marks test knowledge, with an emphasis on the application of knowledge. The remainder test the application of scientific inquiry, analysis and problem solving skills. The first question is usually an extensive data question and there are two extended response questions, one for about 4–5 marks and the other for about 8–10 marks – the longer extended response question will normally have a choice and is usually the last question in the paper.

Altogether, there are 90 marks and you will have 2 hours and 30 minutes to complete the paper. The majority of the marks will be straightforward and linked to grade C but some questions are more demanding and are linked to grade A.

General hints and tips

You should have a copy of the Course Assessment Specification (CAS) for Advanced Higher Biology but, if you haven't got one, make sure to download it from the SQA website. It is worth spending some time looking at this document, as it indicates what you can be tested on in your examination.

This book contains three practice Advanced Higher Biology examination papers. One is the SQA specimen paper, one is a Hodder Gibson model paper and the third is a past exam paper. The model paper has been carefully assembled to be as similar as possible to a typical Advanced Higher Biology examination paper. Notice how similar it is in the way in which it is laid out and the types of question it asks – your own Course examination is going to be very similar as well, so the value of this paper is obvious! Each paper can be attempted in its entirety, or groups of questions on a particular topic or skill area can be attempted. If you are trying a whole examination paper from this book, give yourself a maximum of 2 hours and 30 minutes to complete it. The questions in each paper are laid out roughly in Unit order. Make sure that you spend time in using the answer section to mark your own work – it is especially useful if you can get someone to help you with this.

The marking instructions give acceptable answers with alternatives. You could even grade your work on an A–D basis. The following hints and tips are related to examination techniques as well as avoiding common mistakes. Remember that if you hit problems with a question, you should ask your teacher for help.

Section 1

25 multiple-choice items   **25 marks**

- Answer on a grid.

- Do not spend more than 30 minutes on this section.

- Some individual questions might take longer to answer than others – this is quite normal and make sure you use scrap paper if a calculation or any working is needed.

- Some questions can be answered instantly – again, this is normal.

- Do not leave blanks – complete the grid for each question as you work through.

- Try to answer each question in your head without looking at the options. If your answer is there you are home and dry!

- If you are not certain, it is sometimes best to choose the answer that seemed most attractive on first reading the answer options.

- If you are guessing, try to eliminate options before making your guess. If you can eliminate three, you will be left with the correct answer even if you do not recognise it!

Section 2

Restricted and extended response   **65 marks**

- Spend about 2 hours on this section.

- A clue to your answer length is the mark allocation – questions restricted to 1 mark can be quite short. If there are 2–3 marks available, your answer will need to be extended and may well have two, three or even four parts.

- The questions are usually laid out in Unit sequence but remember that some questions are designed to cover more than one Unit.

- The C-type questions usually start with "State", "Identify", "Give" or "Name" and often need only a single sentence in response. They will usually be for 1 mark each.

- Questions that begin with "Explain", "Suggest" and "Describe" are usually A-type questions and are likely to have more than one part to the full answer. You will usually have to write a sentence or two and there may be 2 or even 3 marks available.

- Make sure you read over the question twice **before** trying to answer – there will be very important information within the question and underlining or highlighting key words is good technique.

- Using abbreviations like DNA and ATP is fine. The Advanced Higher Biology Course Assessment Specification (CAS) will give you the acceptable abbreviations.

- Don't worry if the questions are in unfamiliar contexts, that's the idea! Just keep calm and read the questions carefully.

- In the large data question (Q1), it is good technique to read the whole stem and then skim the data groups before starting to answer any of the parts.

- In the large data question (Q1), be aware that the first piece of data presented should give the main theme of the question.

- In experimental questions, you must be aware of the different classes of variables, why controls are needed and how reliability and validity might be improved. It is worth spending time on these ideas – they are essential and will come up year after year.

- Note that information which is additional to the main stem may be given within a question part – if it's there, you will need it!

- If instructions in the question ask you to refer to specific groups of data, follow these and don't go beyond them.

- Remember that a conclusion can be seen from data, whereas an explanation will usually require you to supply some background knowledge as well.

- Note that in your answer, you may be asked to "use data to…" – it is essential that you do this.

- Remember to "use values from the graph" when describing graphical information in words, if you are asked to do so.

- Look out for graphs with two Y-axes – these need extra special concentration and anyone can make a mistake!

- In numerical answers, it's good technique to show working and supply units.

- Answers to calculations will not usually have more than two decimal places.

- You should round any numerical answers as appropriate, but two decimal places should be acceptable.

- Ensure that you take error bars into account when evaluating the effects of treatments.

- Do not leave blanks. Always have a go, using the language in the question if you can.

Good luck!

Remember that the rewards for passing Advanced Higher Biology are well worth it! Your pass will help you get the future you want for yourself. In the exam, be confident in your own ability. If you're not sure how to answer a question, trust your instincts and just give it a go anyway.

Keep calm and don't panic! GOOD LUCK!

ADVANCED HIGHER

2015 Specimen Question Paper

National
Qualifications
SPECIMEN ONLY

SQ02/AH/02

**Biology
Section 1—Questions**

Date — Not applicable

Duration — 2 hours 30 minutes

Instructions for the completion of Section 1 are given on *Page two* of your question and answer booklet SQ02/AH/01.

Record your answers on the answer grid on *Page three* of your question and answer booklet.

Before leaving the examination room you must give your question and answer booklet to the Invigilator; if you do not, you may lose all the marks for this paper.

SECTION 1 — 25 marks

Attempt ALL questions

1. Which of the following is a covalent bond that stabilises the tertiary structure of a protein?

    A   Disulphide bridge

    B   Hydrogen bond

    C   Ionic bond

    D   Hydrophobic interactions

2. A hydrophobic amino acid has an R group that is

    A   negatively charged

    B   positively charged

    C   not polar

    D   polar.

3. A buffered solution of four amino acids was applied to the midline of a strip of electrophoresis gel. The result of running the gel is shown below.

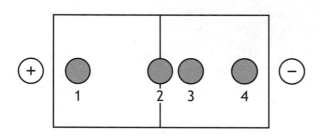

    Which of the amino acids was at its isoelectric point?

    A   1

    B   2

    C   3

    D   4

4. The table shows the number of amino acids in a particular protein and the charge of each amino acid at a certain pH.

Amino acid	Charge	Number
arginine	positive	13
aspartate	negative	9
cysteine	negative	2
histidine	positive	2
glutamate	negative	20
lysine	positive	19
tyrosine	negative	7

Assuming that each amino acid carries a single positive or negative charge, what is the protein's net charge at this pH?

A   −4

B   −38

C   +4

D   +38

5. The diagram below shows how phosphate is used to modify the conformation of an enzyme, phosphorylase, and so change its activity.

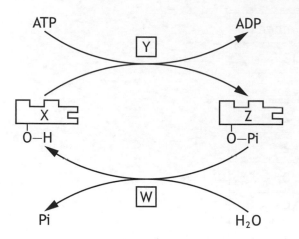

Which line in the table correctly identifies the labels?

	Kinase	Phosphatase	Phosphorylase
A	Y	Z	W
B	W	Y	Z
C	X	Y	W
D	Y	W	Z

6. The diagram below shows the distribution of protein molecules in a cell membrane.

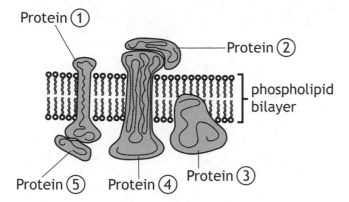

Which line in the table correctly identifies a peripheral and an integral membrane protein?

	Peripheral membrane protein	Integral membrane protein
A	1	5
B	2	1
C	3	4
D	5	2

7. The sodium-potassium pump spans the plasma membrane. Various processes involved in the active transport of sodium and potassium ions take place either inside the cell (intracellular) or outside the cell (extracellular).

Which line in the table correctly applies to the binding of potassium ions?

	Binding location of potassium ions	Conformation of transport protein
A	extracellular	not phosphorylated
B	intracellular	not phosphorylated
C	extracellular	phosphorylated
D	intracellular	phosphorylated

8. The diagram below shows a haemocytometer grid that was used to estimate the number of cells in a $10\,cm^3$ microbial culture. The depth of the counting chamber is $0.2\,mm$.

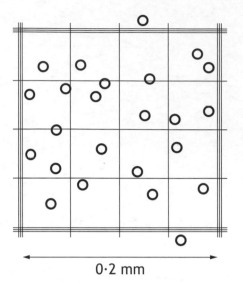

0·2 mm

The number of cells in the $10\,cm^3$ culture was

A 2.75×10^7

B 2.5×10^7

C 2.25×10^7

D 1.6×10^3.

9. The contribution of aquaporins (AQPs) to osmosis was studied by measuring the rate of movement of radioactive water across a plasma membrane. Rates were measured in either isotonic or hypertonic external solution when the pores were either open or closed. Results are shown in the table.

External solution	Rate of water movement (units s^{-1})	
	Open AQPs	Closed AQPs
Isotonic	2·5	1·0
Hypertonic	20·0	1·8

Which of the following is the dependent variable in the experiment?

A   External solution

B   Radioactivity of water

C   Rate of water movement

D   Aquaporins

10. To which group of signalling molecules do steroid hormones belong?

    A   Extracellular hydrophobic

    B   Extracellular hydrophilic

    C   Peptide hormones

    D   Neurotransmitters

11. The following diagrams represent stages in an indirect ELISA used to detect the presence of a poisonous toxin in food samples. The test shown is positive. Identify the diagram that represents the correct sequence of events in the ELISA.

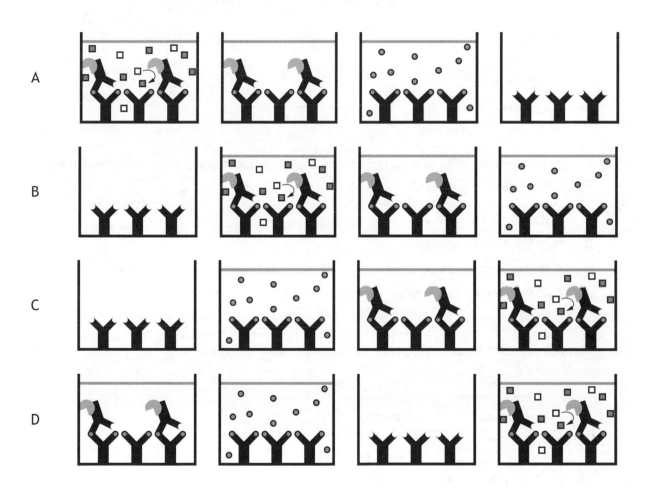

12. Identify which of the following proteins are involved in apoptosis.

    1   Caspases

    2   p53

    3   DNAses

    A   2 only

    B   1 and 2 only

    C   1 and 3 only

    D   1, 2 and 3

13. Animal cells growing in culture are found to spend 20% of their time in the G2 phase of the cell cycle. G2 lasts for 4 hours.

    If cells spend 12% of their time in the M phase, how long does this last?

    A   2 hours 4 minutes

    B   2 hours 12 minutes

    C   2 hours 24 minutes

    D   2 hours 40 minutes

14. Name the ion which is pumped across membranes by bacteriorhodopsin.

    A   Sodium

    B   Potassium

    C   Chloride

    D   Hydrogen

15. Which of the following would be true if a population's gene pool remained unaltered for many generations?

    A   Mating was random

    B   Migration was common

    C   Genetic drift had occurred

    D   Certain alleles had a selective advantage

16. Identify the line in the table that applies to r-selected species.

	many offspring produced	prolonged parental care
A	yes	yes
B	yes	no
C	no	yes
D	no	no

17. *C. elegans* is a model organism of the phylum

    A   Chordata

    B   Arthropoda

    C   Nematoda

    D   Mollusca.

18. From the following list, identify all the possible sources of DNA during horizontal gene transfer.

    1   viruses

    2   plasmids

    3   bacterial cells

    4   gametes

    A   1 and 2 only

    B   2 and 3 only

    C   1, 2 and 3 only

    D   1, 2 ,3 and 4

19. The following diagram is **drawn to scale** and indicates the position of four linked genes on a chromosome.

W   X   Y   Z

Identify the column in the table that gives the correct recombination frequencies for the genes in the chromosome map shown above.

Genes	Recombination frequency (%)			
	A	B	C	D
X and Z	17	19	17	15
W and Z	25	25	23	23
Y and W	19	17	15	17
Z and Y	6	8	8	6
X and W	8	6	6	8

20. The error bars on the graphs represent standard errors in the mean (SEM). Which graph shows significantly different reliable data?

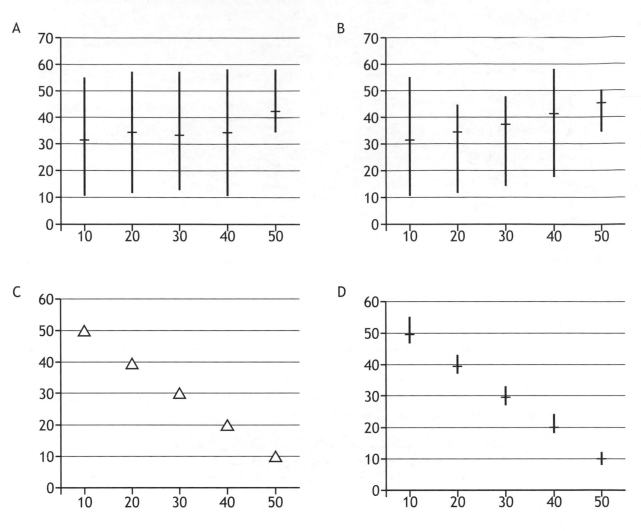

21. *Anolis* lizards are found on Caribbean islands. They feed on prey of various sizes.

    Histogram 1 shows the range of prey length eaten by *Anolis marmoratus* on the island of Jarabacoa, where there are five other *Anolis* species.

    Histogram 2 shows the range of prey length eaten by *Anolis marmoratus* on the island of Marie Galante, where it is the only *Anolis* species.

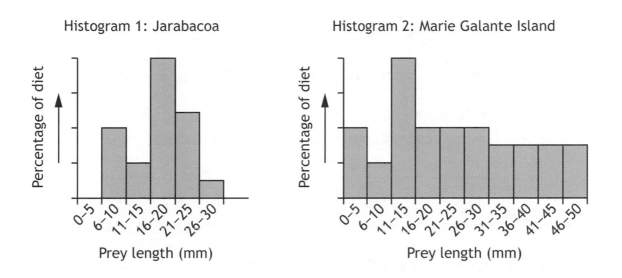

    Which of the following statements could explain the different range of prey sizes eaten by *Anolis marmoratus* on the two islands?

    A   Larger numbers of prey are found on Marie Galante.

    B   *Anolis marmoratus* occupies its fundamental niche on Jarabacoa.

    C   *Anolis marmoratus* occupies its realised niche on Marie Galante.

    D   Resource partitioning takes place on Jarabacoa.

22. Herd immunity threshold is

    A   The density of hosts in a population required to prevent an epidemic

    B   The density of resistant hosts in a population required to prevent an epidemic

    C   The density of hosts in a population required for transmission to cause an epidemic

    D   The density of parasites in a population required to cause an epidemic.

23. Reverse transcriptase catalyses the production of

    A   DNA from RNA

    B   DNA from DNA

    C   mRNA from DNA

    D   tRNA from mRNA.

24. Which of the following would **not** provide long-term control of parasites following a natural disaster?

    A   Immunisation

    B   Improved sanitation

    C   Co-ordinated vector control

    D   Drug treatment of infected humans

25. The formula N = MC/R is used to estimate population size using mark and recapture data.

    N = population estimate

    M = number first captured, marked and released

    C = total number in second capture

    R = number marked in second capture

    In a survey to estimate a woodlouse population, the following data were obtained:

    Woodlice captured, marked and released   = 80

    Marked woodlice in second capture   = 24

    Unmarked woodlice in second capture   = 96

    The estimated population of the woodlice was

    A   200

    B   320

    C   400

    D   3840.

[END OF SECTION 1. NOW ATTEMPT THE QUESTIONS IN SECTION 2 OF YOUR QUESTION AND ANSWER BOOKLET]

AH

National
Qualifications
SPECIMEN ONLY

Mark

SQ02/AH/01

Biology
Section 1 — Answer Grid
and Section 2

Date — Not applicable

Duration — 2 hours 30 minutes

Fill in these boxes and read what is printed below.

Full name of centre

Town

Forename(s)

Surname

Number of seat

Date of birth

Day	Month	Year	Scottish candidate number

Total marks — 90

SECTION 1 — 25 marks

Attempt ALL questions.

Instructions for completion of Section 1 are given on *Page two*.

SECTION 2 — 65 marks

Attempt ALL questions.

Write your answers clearly in the spaces provided in this booklet. Additional space for answers and rough work is provided at the end of this booklet. If you use this space you must clearly identify the question number you are attempting. Any rough work must be written in this booklet. You should score through your rough work when you have written your final copy.

Use **blue** or **black** ink.

Before leaving the examination room you must give this booklet to the Invigilator; if you do not you may lose all the marks for this paper.

SECTION 1 — 25 marks

The questions for Section 1 are contained in the question paper SQ02/AH/02.
Read these and record your answers on the answer grid on *Page three* opposite.
Use **blue** or **black** ink. Do NOT use gel pens or pencil.

1.   The answer to each question is **either** A, B, C or D. Decide what your answer is, then fill in the appropriate bubble (see sample question below).

2.   There is **only one correct** answer to each question.

3.   Any rough working should be done on the additional space for answers and rough work at the end of this booklet.

Sample Question

The thigh bone is called the

     A   humerus

     B   femur

     C   tibia

     D   fibula.

The correct answer is **B**—femur. The answer **B** bubble has been clearly filled in (see below).

Changing an answer

If you decide to change your answer, cancel your first answer by putting a cross through it (see below) and fill in the answer you want. The answer below has been changed to **D**.

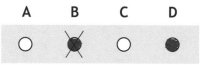

If you then decide to change back to an answer you have already scored out, put a tick (✓) to the **right** of the answer you want, as shown below:

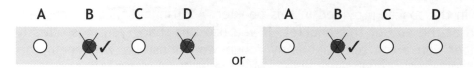

SECTION 1 — Answer Grid

	A	B	C	D
1	○	○	○	○
2	○	○	○	○
3	○	○	○	○
4	○	○	○	○
5	○	○	○	○
6	○	○	○	○
7	○	○	○	○
8	○	○	○	○
9	○	○	○	○
10	○	○	○	○
11	○	○	○	○
12	○	○	○	○
13	○	○	○	○
14	○	○	○	○
15	○	○	○	○
16	○	○	○	○
17	○	○	○	○
18	○	○	○	○
19	○	○	○	○
20	○	○	○	○
21	○	○	○	○
22	○	○	○	○
23	○	○	○	○
24	○	○	○	○
25	○	○	○	○

SECTION 2 — 65 marks

Attempt ALL questions

It should be noted that question 11 contains a choice.

1. Recently a new class of RNA, called microRNA, has been discovered. These small RNA molecules have an important role in controlling the translation of mRNA. This type of control is called RNA interference.

   A microRNA is formed from a precursor RNA molecule that folds into a double-stranded "hairpin" structure. The hairpin is then processed to give a shorter molecule by the enzymes "Drosha" and "Dicer". One strand of this short molecule attaches to RISC proteins; the resulting complex binds to target mRNA molecules and prevents translation (Figure 1).

Figure 1: Control of gene expression by RNA interference

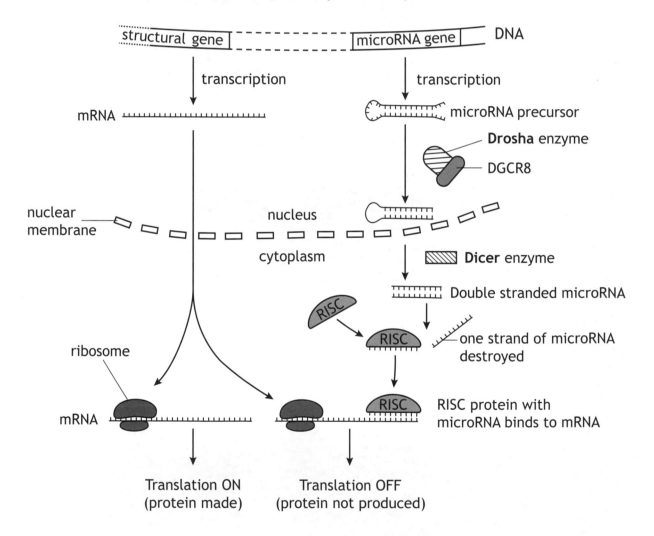

Recent research has investigated the importance of microRNA in controlling the fate of stem cells. Stem cells can either divide rapidly to make more stem cells, a process called **self-renewal**, or differentiate into specialised cell types. To determine the role of microRNAs in these processes, stem cells were modified to "knock out" microRNA production. These microRNA knockout cells lack the protein DGCR8, an activator of Drosha. Figures 2A and 2B compare growth rate and cell-cycle progression in knockout and normal cells.

1.   (continued)

In further work, the differentiation of knockout and normal cells was studied by inducing the cells to differentiate. Analysis was carried out on the levels of specific marker molecules whose presence is associated with either self-renewal or differentiation. Results are shown in Figures 3A and 3B.

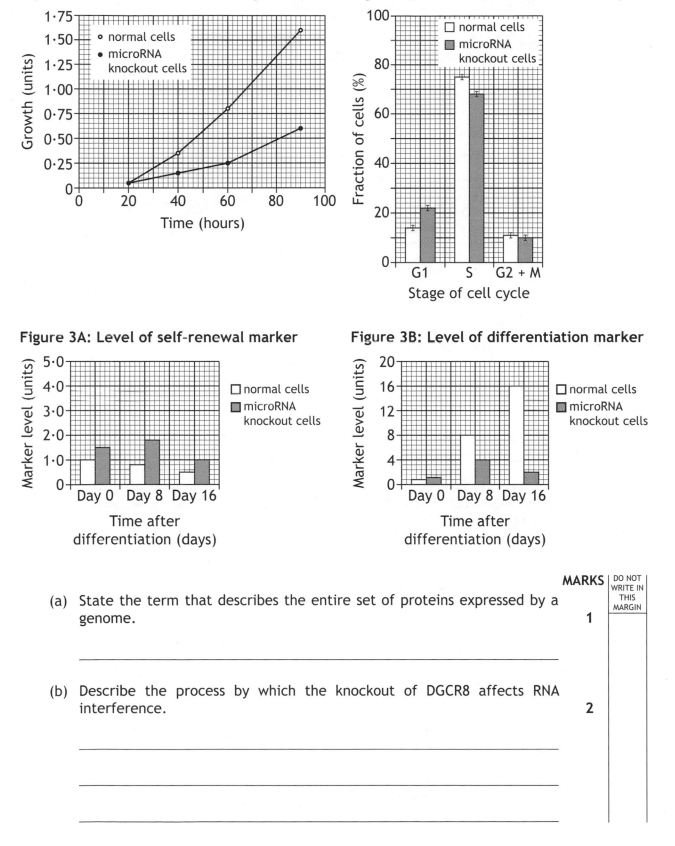

Figure 2A: Effect of knockout on growth rate

Figure 2B: Effect of knockout on cell cycle

Figure 3A: Level of self-renewal marker

Figure 3B: Level of differentiation marker

MARKS

(a)   State the term that describes the entire set of proteins expressed by a genome.

1

_____

(b)   Describe the process by which the knockout of DGCR8 affects RNA interference.

2

_____

_____

_____

MARKS | DO NOT WRITE IN THIS MARGIN

1.   (continued)

(c)   (i)   Describe what happens during the G1 phase of the cell cycle.   **1**

_____

_____

(ii)   Using Figure 2A, calculate the percentage reduction in growth at 90 hours caused by the microRNA knockout.   **1**

_____

_____

(iii)   The researchers concluded that microRNA knockout cells do not progress normally through the cell cycle.

Explain how the results in Figure 2B support this conclusion.   **2**

_____

_____

_____

(d)   (i)   Use Figures 3A and 3B. Give one general conclusion about the expression of the differentiation marker by comparing normal and knockout cells.   **1**

_____

_____

(ii)   There is a hypothesis that self-renewal is switched off as differentiation proceeds and that the interaction of these two processes is abnormal in knockout cells.

Explain how the data support this hypothesis.   **2**

_____

_____

_____

MARKS | DO NOT WRITE IN THIS MARGIN

2. Gamma-aminobutyric acid (GABA) is a neurotransmitter that functions as a signalling molecule in the central nervous system. GABA binds to a receptor protein located in the plasma membrane of target cells as shown in Figure 1. Binding of a GABA molecule opens a channel that allows chloride ions (Cl^-) to enter the cell.

Figure 1

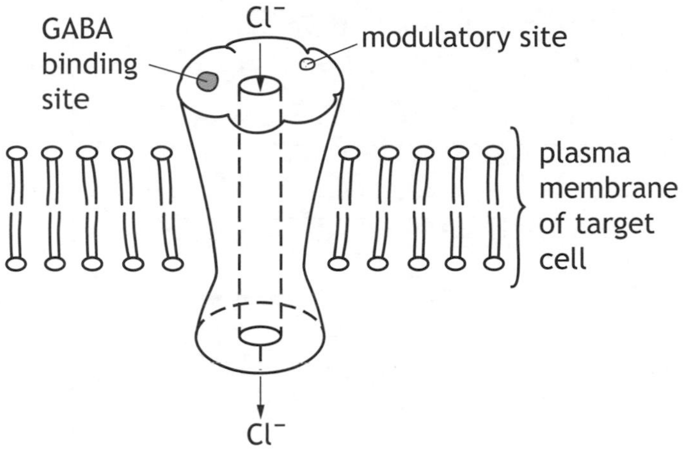

Figure 2

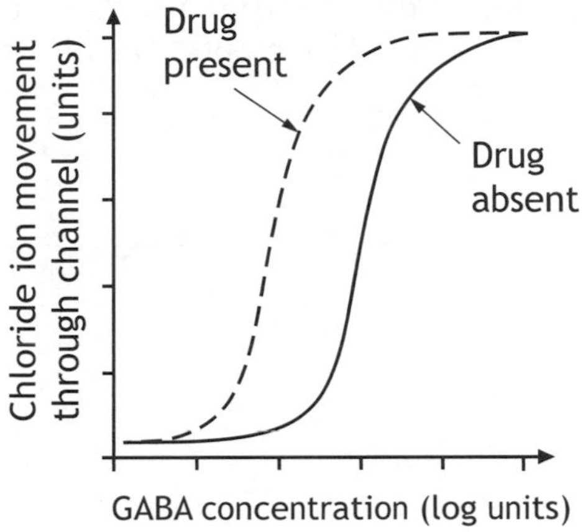

Benzodiazepines are sedative drugs that bind to the receptor protein and increase its affinity for GABA. These drugs act as allosteric modulators by binding at a site that is distinct from the GABA-binding site. Figure 2, above, shows the movement of chloride ions through the channel as GABA is increased with and without the drug being present.

(a) Using the information provided, explain why the GABA receptor is described as a ligand-gated channel.   **2**

(b) State the term that describes the action of a membrane receptor in which signal binding brings about an effect in the cytoplasm.   **1**

(c) (i) Describe the information in Figure 2 that shows that the affinity of the receptor for GABA has been increased by the benzodiazepine.   **1**

MARKS | DO NOT WRITE IN THIS MARGIN

2. (continued)

(ii) Explain why the affinity of the receptor for GABA increases when the drug binds to the modulatory site.

1

_____

_____

(iii) Describe the effect that chloride ion influx will have on the membrane potential of the nerve cell.

1

_____

_____

MARKS | DO NOT WRITE IN THIS MARGIN

3. An investigation into the effects of different concentrations of ATP on muscle tissue used muscle from three pork chops (A, B and C), all bought from the same shop.

Three thin strips of muscle were cut from each chop and placed on microscope slides. The length of each strip was measured and recorded.

Equal volumes of a 10% ATP solution were added to one strip of muscle from each chop and the length of each measured again.

The experiment was repeated using a 5% ATP solution on the second set of strips and distilled water on the final set.

(a) Identify the independent variable in this experiment.   1

_____

(b) Two confounding variables in this experiment are temperatures of the solutions and muscle strips during the experiment, and the breed of pig that the chops came from.

(i) Suggest one further confounding variable in this experiment.   1

_____

_____

(ii) Explain the way in which this variable could affect the outcome of this experiment.   1

_____

_____

The following table shows the data collected.

Solution added to strip	Pork chop strip sample	Initial length (mm)	Final length (mm)	Change in length (mm)
10% ATP	A	10	8	2
	B	11	8	3
	C	10	11	1
5% ATP	A	12	11	1
	B	13	12	1
	C	11	10	1
Distilled water	A	12	12	0
	B	12	13	1
	C	9	10	1

MARKS | DO NOT WRITE IN THIS MARGIN

3. **(continued)**

(c) State whether or not the data is reliable. Explain your answer. 2

_____

_____

_____

(d) Name the type of control used in this experiment. 1

_____

(e) Suggest how selection bias has affected the validity of this experiment. 1

_____

_____

MARKS | DO NOT WRITE IN THIS MARGIN

4. When insulin attaches to its receptor in the plasma membrane of fat cells and muscle cells, GLUT 4 glucose transporter proteins in the cytoplasm are recruited into the membrane to take in glucose. Type 2 diabetes is associated with insulin resistance in which cells are less able to respond to insulin in this way.

A recent study concluded that moderate strength training increases the GLUT 4 content of muscle tissue in those with type 2 diabetes. Individuals taking part all did strength training on one leg (T leg) for six weeks while the other leg was left untrained (UT leg). The subjects either had type 2 diabetes or did not. At the end of the training, muscle biopsies (samples) were taken from the trained and untrained legs and compared for GLUT 4 protein content. The results are shown in the graph below.

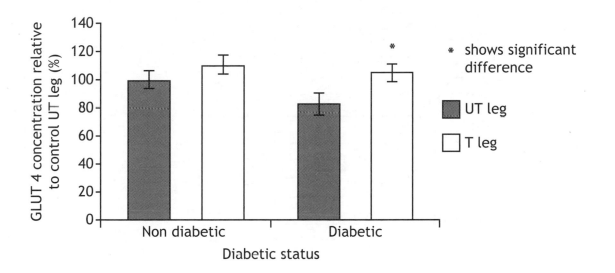

(a) State a suitable null hypothesis in this investigation. 1

_____

_____

(b) The researchers concluded that moderate strength training increases the GLUT 4 content of muscle tissue **only** in those with type 2 diabetes. Identify the evidence that supports this conclusion. 2

_____

_____

_____

(c) State why the treatment regimes for subjects with type 1 diabetes may differ from subjects with type 2 diabetes. 1

_____

_____

MARKS | DO NOT WRITE IN THIS MARGIN

5. Thyroxine is a hormone that acts as a regulator of metabolic rate in most tissues. Thyroxine causes an increase in metabolic rate by binding to specific receptors located within the nucleus of a target cell. Hyperthyroidism is a condition caused by overproduction of thyroxine. The following graph shows the average change in metabolic rate of individuals with hyperthyroidism who were treated over a 20-week period with a drug (carbimazole). The drug decreases the synthesis of thyroxine from the thyroid gland.

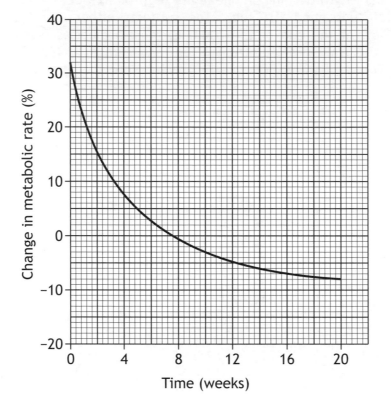

(a) State the property of thyroxine that allows it to cross the membrane of cells.

1

_____

(b) Describe the mechanism by which thyroxine binding to its receptor affects transcription of genes that influence metabolic rate.

2

_____

_____

_____

MARKS | DO NOT WRITE IN THIS MARGIN

5.   (continued)

(c)   (i)   Explain how the data support the conclusion that the thyroid gland has large stores of thyroxine.

1

_____

_____

(ii)   Explain why the changes in metabolic rate have been presented as percentages.

1

_____

_____

MARKS | DO NOT WRITE IN THIS MARGIN

6. Rod cells and cone cells are photoreceptors in vertebrate eyes. Membranes in these cells contain rhodopsin, a protein molecule that has a light-absorbing component. Rhodopsin generates a nerve impulse when light is absorbed.

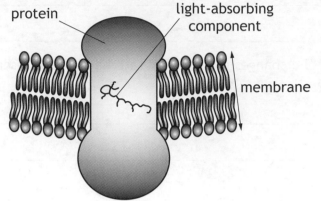

protein   light-absorbing component

membrane

(a) Name the light-absorbing component of rhodopsin.

1

_____

(b) Explain the mechanism by which the absorption of a photon by rhodopsin leads to the generation of a nerve impulse.

2

_____

_____

_____

_____

(c) Give one feature of the photoreceptor system in rods that allows these cells to function in low light intensity.

1

_____

_____

MARKS | DO NOT WRITE IN THIS MARGIN

7. A type of haemophilia results when a gene that codes for a blood clotting factor, factor VIII, is mutated. This gene is located on the X chromosome. Mutated alleles do not produce functional factor VIII.

(a) Explain why men are more likely than women to be affected by this type of haemophilia.

2

_____

_____

_____

(b) An unaffected man and a carrier woman have a daughter and a son.

State the probability of each child being able to produce functional factor VIII.

2

Space for calculation and working

Daughter _____

Son   _____

(c)   (i) Explain the importance of inactivation of the X chromosome in females.

1

_____

_____

(ii) Analysis of a female carrier showed that her blood contained only 42% of the normal levels of functional factor VIII.

Suggest why this value was lower than predicted.

1

_____

_____

MARKS | DO NOT WRITE IN THIS MARGIN

8. Describe how the events that occur during crossing over contribute to the production of variable gametes.

4

_____

_____

_____

_____

_____

_____

9. The following figure shows the life cycle of the macroparasitic flatworm called *Schistosoma japonicum*. The flatworm can live for many years within a host. In humans, if untreated, it causes the disease schistosomiasis (bilharzia) that can be fatal.

Life cycle of *Schistosoma japonicum*

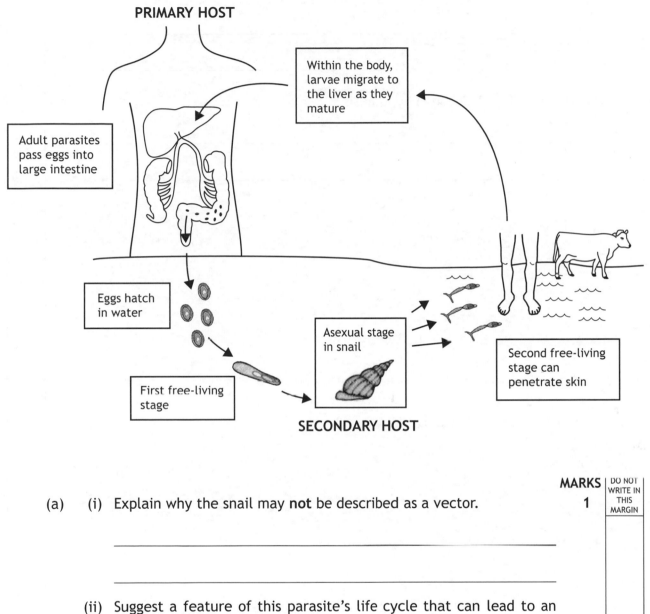

	MARKS	DO NOT WRITE IN THIS MARGIN

(a) (i) Explain why the snail may **not** be described as a vector.　　　　1

_____

_____

(ii) Suggest a feature of this parasite's life cycle that can lead to an increased rate of transmission.　　　　1

_____

_____

MARKS | DO NOT WRITE IN THIS MARGIN

9. (continued)

(b) Parasites living inside a host will be exposed to attack by the host's immune system.

Describe one way in which parasites may overcome the immune response of their hosts.

1

_____

_____

(c) Describe the Red Queen hypothesis.

2

_____

_____

MARKS | DO NOT WRITE IN THIS MARGIN

10. Fur seals spend most of their lives feeding in Antarctic seas. During the short summer they come ashore to breed.

The figure below shows the number of fur seals breeding on Signy Island from 1956 to 1986.

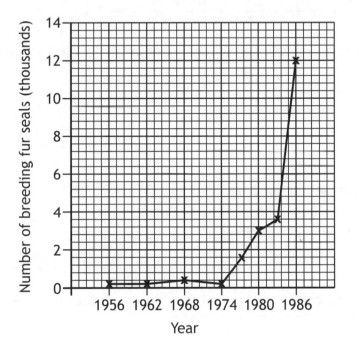

(a) Calculate the percentage increase in the size of the breeding seal population between 1980 and 1986.

Space for calculation and working

1

_____ %

MARKS | DO NOT WRITE IN THIS MARGIN

10.   (continued)

(b) Permanent quadrats were established to investigate the effect of fur seals on ground cover plants. The table shows the mean percentage of cover of a number of plant species sampled in the permanent quadrats in 1965 and 1985.

Plant species	Percentage cover (%)	
	1965	1985
Drepanocladus uncinatus	30	0
Bryum algens	49	0
Tortula filaris	16	0
Tortula saxicola	4	4
Prasiola crispa	1	41

(i) Explain the changes in percentage cover between 1965 and 1985.   2

_____

_____

_____

(ii) Suggest why the percentage cover in 1985 is not 100%.   1

_____

_____

(c)   (i) Describe one consideration that must be taken into account when carrying out sampling in an ecosystem.   1

_____

_____

(ii) Describe the process of stratified sampling.   1

_____

_____

MARKS | DO NOT WRITE IN THIS MARGIN

11. Answer **either A or B** in the space below.

A   Describe the specific cellular defences that protect mammals from parasite infection.   **8**

OR

B   Describe courtship behaviours that affect reproductive success.   **8**

[END OF SPECIMEN QUESTION PAPER]

ADDITIONAL SPACE FOR ANSWERS AND ROUGH WORK

MARKS | DO NOT WRITE IN THIS MARGIN

ADDITIONAL SPACE FOR ANSWERS AND ROUGH WORK

[BLANK PAGE]

DO NOT WRITE ON THIS PAGE

ADVANCED HIGHER

Model Paper

Whilst this Model Paper has been specially commissioned by Hodder Gibson for use as practice for the Advanced Higher (for Curriculum for Excellence) exams, the key reference document remains the SQA Specimen Paper 2015 and SQA Past Paper 2016.

National
Qualifications
MODEL PAPER

Biology
Section 1—Questions

Duration — 2 hours 30 minutes

Instructions for the completion of Section 1 are given on *Page two* of your question and answer booklet.

Record your answers on the answer grid on *Page three* of your question and answer booklet.

Before leaving the examination room you must give your question and answer booklet to the Invigilator; if you do not, you may lose all the marks for this paper.

SECTION 1 — 25 marks

Attempt ALL questions

1. Immobilised metal ion affinity chromatography (IMAC) can be used to purify proteins. This technique works by allowing proteins with an affinity for metal ions to be retained in a column containing immobilised metal ions, such as cobalt.

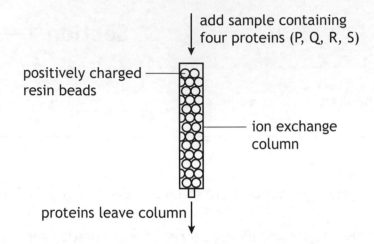

add sample containing four proteins (P, Q, R, S)

positively charged resin beads

ion exchange column

proteins leave column

The more negatively charged the protein, the longer it takes to pass through the column.

The graph below shows the time taken for the four proteins to leave the column.

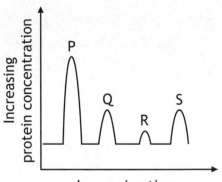

Increasing protein concentration

Increasing time

Which conclusion can be drawn from this experiment?

A   Protein P is more negatively charged than protein R.

B   Protein S is more negatively charged than protein Q.

C   Protein R is less negatively charged than protein Q.

D   Protein S is less negatively charged than protein R.

2. The enzyme Na/KATPase moves ions across membranes in the ratio 3 sodium : 2 potassium. 5000 of these ions are pumped across a membrane every 10 seconds.

   The number of potassium ions moved across this membrane in one second is

   A   200

   B   500

   C   2000

   D   3000.

3. The diagram below represents a plastic well from an immunoassay kit, testing a blood sample from a person who had been exposed to a particular virus. The substrate has been broken down to form a coloured product, so the result is positive.

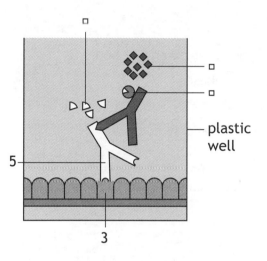

   Which line in the table below identifies the numbered components in the well?

	Antigen	Antibody	Reporter enzyme	Substrate
A	3	5	2	4
B	4	1	2	5
C	3	5	4	1
D	5	2	1	4

4. Which of the following diagrams represents the sequence of phases in the cell cycle?

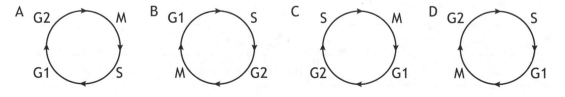

5. The diagram below shows the changes in cell mass and DNA mass during two cell cycles.

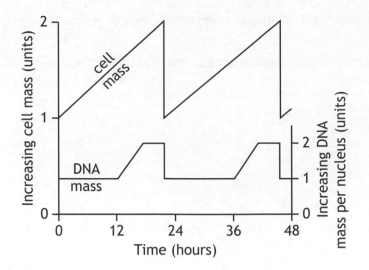

Which valid conclusion could be drawn from the graph?

During the cell cycle

A   replication takes place between 10 and 12 hours

B   mitosis is divided into four phases

C   interphase is the longest phase

D   cytokinesis takes place at 12 and 36 hours.

6. Which of the following techniques can be used to estimate the total cell count?

A   immunoassay

B   haemocytometry

C   chromatography

D   electrophoresis

7. Retinoblastoma protein (Rb) has a role in the regulation of progress through the cell cycle. It can be phosphorylated (Rb-P) or not phosphorylated (Rb).

Which line in the table below shows the phase in the cell cycle during which Rb functions as a regulator and which of its phosphorylation states allows the cell cycle to progress?

	Phase	*Phosphorylation state which allows the cycle to progress*
A	G1	phosphorylated
B	G1	not phosphorylated
C	S	phosphorylated
D	S	not phosphorylated

8. Which line in the table below identifies factors which trigger apoptosis?

	p53 protein	Cell growth factors
A	present	absent
B	present	present
C	absent	absent
D	absent	present

9. Colorimetry was used to produce the standard curve for soluble protein concentration shown below.

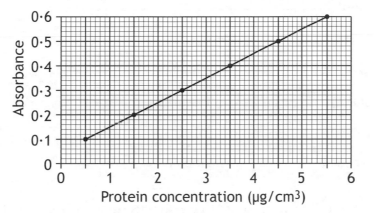

In an experiment to find the soluble protein content of potato tubers, 25 g of fresh potato tissue was ground with 50 cm³ of buffer and centrifuged. A total of 65 cm³ of extract was produced.

1 cm³ of the extract was tested in a colorimeter and gave an absorbance of 0·5.

What was the total soluble protein content of the fresh potato tissue in µg per g?

A   3·9

B   9·0

C   11·7

D   13·5

10. The sodium—potassium pump's mechanism of action involves the stages shown below.

P   membrane protein is phosphorylated

Q   sodium ions bind to membrane protein

R   sodium ions are released

S   membrane protein changes conformation

The correct sequence of stages in the action of the pump is

A   P, Q, R, S

B   Q, P, S, R

C   Q, P, R, S

D   P, Q, S, R.

11. The diagram below shows a structure from a eukaryotic cell about to undergo cell division.

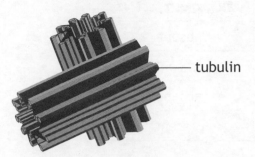

— tubulin

Which line in the table below shows the name of the structure and the protein group to which tubulin belongs?

	Name of structure	Protein group
A	cytoskeleton	globular
B	cytoskeleton	integral
C	centrosome	integral
D	centrosome	globular

12. Which line in the table below represents the binding site and effect on affinity of an allosteric enzyme binding with a positive modulator?

	Modulator binding site		Affinity of enzyme for substrate	
	Active site	Secondary site	Increased	Decreased
A	✓		✓	
B		✓		✓
C		✓	✓	
D	✓			✓

13. The list below contains terms which refer to signalling molecules.

1   hydrophobic

2   hydrophilic

3   steroid

4   peptide

Which of the terms describe the hormone insulin?

A   1 and 3 only

B   1 and 4 only

C   2 and 3 only

D   2 and 4 only

14. Cortisol is a steroid hormone.

Which letter in the diagram below shows movement by molecules of this hormone in the first stage of its cell signalling process?

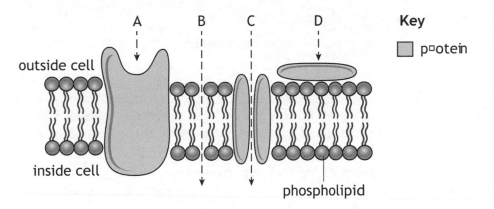

15. The fundamental niche of a species

A   includes the set of resources available in the absence of competition

B   includes the set of resources available in the presence of competition

C   permits co-existence in a community

D   permits the sharing of resources with other species.

16. The regions of the graph below indicate conditions in four habitats.

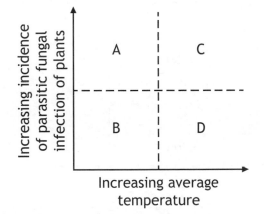

In which habitat is parthenogenesis most likely to evolve as a successful reproductive strategy?

17. Crossing over that generates new allele combinations during meiosis occurs between

    A   sister chromatids of homologous chromosomes

    B   non-sister chromatids of homologous chromosomes

    C   sister chromatids of non-homologous chromosomes

    D   non-sister chromatids of non-homologous chromosomes.

18. A quadrat with sides 50 cm long was used to estimate the densities of the barnacle *Balanus balanoides* in two areas, X and Y, on a rocky shore. Five random samples were taken in each of the two areas and the results are given in the table below.

Quadrat number	Number of individual barnacles	
	Area X	Area Y
1	270	150
2	190	160
3	390	420
4	190	310
5	110	160

Which line in the table below shows the mean density per square metre in the two areas?

	Area X	Area Y
A	230	150
B	460	480
C	920	960
D	1 150	1 200

19. Independent assortment of chromosomes during meiosis results in the production of gametes with varied combinations of chromosomes.

How many different combinations of chromosomes are possible in the gametes of an organism with a haploid number of 3?

    A   4

    B   6

    C   8

    D   12

20. *Dicrocoelium dendriticum* is a flatworm parasite of grazing animals, such as sheep and cattle.

Which line in the table shows the phyla to which these species belong?

	Dicrocoelium	*Cattle and sheep*
A	nematoda	chordata
B	platyhelminthes	arthropoda
C	nematoda	arthropoda
D	platyhelminthes	chordata

21. The beef tapeworm *Taenia saginata* is a parasite which does not have a digestive system during part of its life cycle.

For this reason, the parasite is described as

A degenerate

B being an ectoparasite

C occupying its fundamental niche

D co-existing by resource partitioning.

22. The virulence of an infectious organism is defined as the case fatality risk (CFR). CFR can be defined as the percentage of infections which result in death. The table below shows the numbers of people infected by "bird flu" virus (H5N1) in an area and the numbers who died from it over a five year period.

Year	2004	2005	2006	2007	2008
Total infections of H5N1	46	98	115	88	44
Total number of deaths from H5N1 infections	32	43	79	59	33

In which year was H5N1 most virulent in this area?

A 2004

B 2006

C 2007

D 2008

23. Measles vaccinations are given to as many children as possible in Scotland. This helps to prevent the spread of measles and gives some protection to non-vaccinated and vulnerable children.

    The protection of non-vaccinated, vulnerable children in this instance is an example of

    A   herd immunity   B   epidemiology

    C   immune surveillance   D   immunological memory.

24. The diagram below shows part of the infection cycle of a human T-cell by an HIV retrovirus.

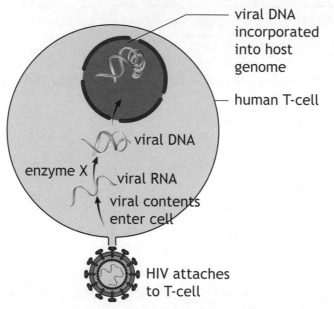

    Enzyme X is

    A   DNAase

    B   DNA polymerase

    C   RNA polymerase

    D   reverse transcriptase.

25. The information in the table below explains terms used in biological science investigations.

    Which line in the table is **not** correct?

	Term	Explanation
A	pilot study	guides modification of experimental design
B	hypothesis	proposes an association between the independent and dependent variable
C	confidence interval	indicates the variability of the data around a mean
D	positive control	provides results in the absence of the treatment

**[END OF SECTION 1. NOW ATTEMPT THE QUESTIONS IN SECTION 2
OF YOUR QUESTION AND ANSWER BOOKLET]**

AH

National
Qualifications
MODEL PAPER

Mark

Biology
Section 1 — Answer Grid
and Section 2

Duration — 2 hours 30 minutes

Fill in these boxes and read what is printed below.

Full name of centre

Town

Forename(s)

Surname

Number of seat

Date of birth

Day	Month	Year	Scottish candidate number

Total marks — 90

SECTION 1 — 25 marks

Attempt ALL questions.

Instructions for completion of Section 1 are given on *Page two*.

SECTION 2 — 65 marks

Attempt ALL questions.

Write your answers clearly in the spaces provided in this booklet. Additional space for answers and rough work is provided at the end of this booklet. If you use this space you must clearly identify the question number you are attempting. Any rough work must be written in this booklet. You should score through your rough work when you have written your final copy.

Use **blue** or **black** ink.

Before leaving the examination room you must give this booklet to the Invigilator; if you do not you may lose all the marks for this paper.

SECTION 1— 25 marks

The questions for Section 1 are contained on *Pages 48–56*.
Read these and record your answers on the answer grid on *Page 59*.
Use **blue** or **black** ink. Do NOT use gel pens or pencil.

1. The answer to each question is **either** A, B, C or D. Decide what your answer is, then fill in the appropriate bubble (see sample question below).

2. There is **only one correct** answer to each question.

3. Any rough working should be done on the additional space for answers and rough work at the end of this booklet.

Sample Question

The thigh bone is called the

   A   humerus

   B   femur

   C   tibia

   D   fibula.

The correct answer is **B**—femur. The answer **B** bubble has been clearly filled in (see below).

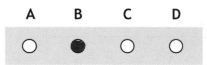

Changing an answer

If you decide to change your answer, cancel your first answer by putting a cross through it (see below) and fill in the answer you want. The answer below has been changed to **D**.

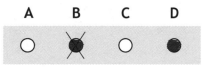

If you then decide to change back to an answer you have already scored out, put a tick (✓) to the **right** of the answer you want, as shown below:

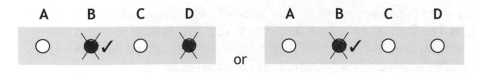

SECTION 1 — Answer Grid

	A	B	C	D
1	◯	◯	◯	◯
2	◯	◯	◯	◯
3	◯	◯	◯	◯
4	◯	◯	◯	◯
5	◯	◯	◯	◯
6	◯	◯	◯	◯
7	◯	◯	◯	◯
8	◯	◯	◯	◯
9	◯	◯	◯	◯
10	◯	◯	◯	◯
11	◯	◯	◯	◯
12	◯	◯	◯	◯
13	◯	◯	◯	◯
14	◯	◯	◯	◯
15	◯	◯	◯	◯
16	◯	◯	◯	◯
17	◯	◯	◯	◯
18	◯	◯	◯	◯
19	◯	◯	◯	◯
20	◯	◯	◯	◯
21	◯	◯	◯	◯
22	◯	◯	◯	◯
23	◯	◯	◯	◯
24	◯	◯	◯	◯
25	◯	◯	◯	◯

SECTION 2 — 65 marks

Attempt ALL questions

It should be noted that question 11 contains a choice.

1. Muscle tissue contains both red and white muscle cell types. Glucose is transported into these cells through trans-membrane proteins called glucose transporters (GLUTs).

    GLUT1 is responsible for glucose uptake in all cells. The membranes of muscle cells also contain GLUT4.

    An investigation was carried out into how the two different GLUTs contributed to glucose uptake by the two different types of muscle cell before and after exposure to insulin. The results are shown in **Figure 1**.

Figure 1: Glucose transport by cells with and without insulin

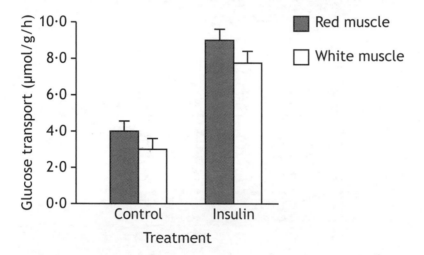

Membranes from muscle cells were isolated and centrifuged to separate the plasma membranes (PM) from internal membranes (IM). The protein components of the membranes were separated by gel electrophoresis and then blotted. The resulting blots were exposed to radioactively labelled antibodies specific to each of the two GLUT proteins to allow identification and quantification. **Figure 2** shows the effects of exposure to insulin on the GLUT levels in each type of muscle.

Figure 2: Effect of insulin on total GLUT levels in muscle cells

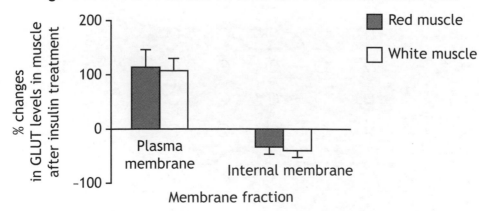

MARKS | DO NOT WRITE IN THIS MARGIN

1. (continued)

In **Figure 3** the size and darkness of the blots indicates the levels of the two GLUTs in the membranes of red and white muscle cells.

Figure 3

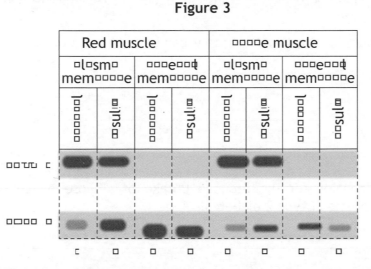

(a) Use data from **Figure 1** to support the statement that red muscle cells have a greater capacity for glucose transport than white muscle cells.   **2**

_____

_____

_____

(b) From the results shown in **Figure 2**, it was concluded that both muscle types have the same underlying GLUT level response to insulin.

Explain how the bars of standard error support this conclusion.   **1**

_____

_____

(c) Using data from **Figure 3**:

(i) describe the distribution of GLUT1 in muscle cells before insulin treatment;   **1**

_____

_____

(ii) give one conclusion about the effect of insulin treatment on GLUT1;   **1**

_____

(iii) give evidence that the effect of insulin on GLUT4 is the same in both types of muscle cell.   **1**

_____

_____

MARKS

1. **(continued)**

(d) A hypothesis was proposed which suggested that insulin triggers the transport of GLUT4 to the plasma membrane from the internal membranes and that more of this transport occurs in red muscle cells compared with white.

Outline the support for this hypothesis that can be seen in **Figure 3**.

2

_____

_____

_____

(e) Suggest an explanation for the reduced uptake of glucose by cells, which is characteristic of type 2 diabetes.

2

_____

_____

_____

MARKS | DO NOT WRITE IN THIS MARGIN

2. The diagram below shows events which occur in part of the thylakoid membrane of a chloroplast from a green plant after a photon of light strikes a chlorophyll molecule.

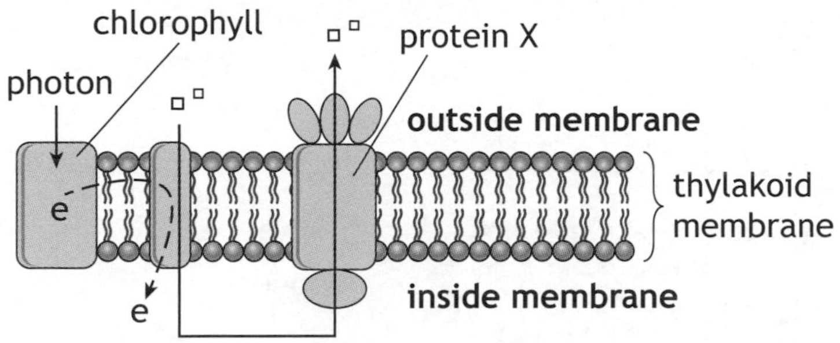

(a) Describe the mechanisms by which:

    (i) absorption of a photon by chlorophyll results in the movement of H^+ ions to the inside of the thylakoid membrane;   **2**

_____

_____

_____

    (ii) the H^+ ions return through the thylakoid membrane to the outside.   **1**

_____

_____

(b) Name protein **X** through which the H^+ ions pass and explain the advantage of its action to the plant.   **2**

_____

_____

_____

MARKS | DO NOT WRITE IN THIS MARGIN

3. The phospholipid layer in membranes acts as a barrier to some molecules, although others can pass through. Transmembrane proteins can act as channels or transporters to perform specific functions.

The diagram shows three transmembrane proteins, X, Y and Z, in a phospholipid membrane.

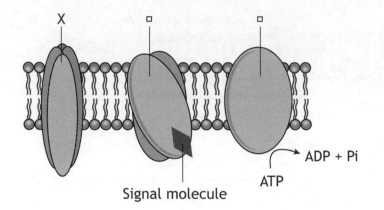

(a) Name a substance whose molecules can pass through the phospholipid bilayer.

1

_____

_____

(b) (i) Protein **X** represents aquaporin.

Name the substance this protein transports and describe how its molecules move through the channel.

2

_____

_____

_____

(ii) Protein **Y** is a gated channel controlled by a signal molecule as shown.

Give one other type of gated channel which can occur in phospholipid membranes.

1

_____

_____

(iii) State the role of ATP in the action of transporter protein **Z**.

1

_____

_____

MARKS | DO NOT WRITE IN THIS MARGIN

4. The chart below describes stages in the production of monoclonal antibody.

Stage description	Diagram
Stage 1 B lymphocytes obtained from treated mice and fusion with mouse melanoma cells attempted	
Stage 2 Resulting cells transferred to plastic wells containing selective medium and supernatant fluid screened for presence of desired antibody	
Stage 3 Cells selected from wells with desired antibody isolated and cloned	
Stage 4 Desired antibody isolated and used in medical procedures	

(a) Describe how mice would be treated prior to **Stage 1** so that they would produce the required B lymphocytes.

1

_____

_____

(b) Explain why the mouse B lymphocytes must be hybridised with melanoma cells in **Stage 1**.

1

_____

_____

(c) Give **two** uses of monoclonal antibodies in medical procedures.

2

_____

_____

_____

MARKS | DO NOT WRITE IN THIS MARGIN

5. In multicellular organisms, the process of apoptosis in target cells is triggered by cell death signals, which may originate within or outwith the cell. The flow chart below shows the events triggered by a cell death signal from outwith the cell.

| Cell death signal molecules released from a lymphocyte | → | Signal molecule binds to surface receptor on target cell | → | Degrading enzymes in target cell activated | → | Target cell destroyed |

(a) Give **one** reason why programmed cell death initiated by a lymphocyte can be beneficial to multicellular organisms.

1

_____

_____

(b) (i) Name **two** degrading enzymes activated during apoptosis.

2

_____

_____

(ii) Bcl-2 is a regulator protein which can inhibit apoptosis. In humans this protein is encoded by the BCL2 gene. Mutations in this gene can increase the levels of Bcl-2.

Suggest why these mutations are usually associated with tumour growth.

2

_____

_____

_____

(c) Give **one** example of an event originating within a cell which can trigger apoptosis.

1

_____

_____

MARKS | DO NOT WRITE IN THIS MARGIN

6. Describe non-specific defences against disease in mammals. 4

MARKS | DO NOT WRITE IN THIS MARGIN

7. The ruff, *Philomachus pugnax*, is a medium-sized ground-nesting wading bird. Males display and fight for females during the breeding season at a grassy site called a lek. There are two male forms: territorial males with conspicuous dark plumage and satellite males with conspicuous light plumage. Territorial males occupy and defend the best mating territories in the lek. Satellite males don't hold territory but enter leks and attempt to mate with females visiting those territories occupied by territorial males. The presence of both types of male in a territory attracts additional females. Females are polygamous and lack the conspicuous plumage of males.

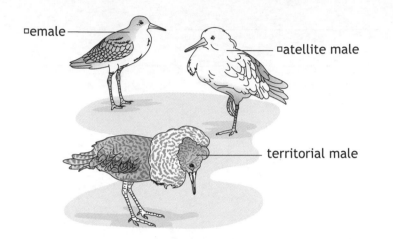

(a) State the term used to describe structural differences between males and females of the same species.   1

_____

(b) Explain the selective advantages which each type of male ruff gains in securing mating opportunities with females.

   (i) Territorial male   1

   _____

   _____

   (ii) Satellite male   1

   _____

   _____

(c) (i) Describe what is meant by the term polygamous.   1

   _____

   _____

   (ii) Explain the advantage to female ruff of having inconspicuous plumage.   1

   _____

   _____

MARKS | DO NOT WRITE IN THIS MARGIN

8. *Prestonella bowkeri* is a small terrestrial snail found in rocky cliff face habitats on the Great Escarpment of southern Africa. The species lives in cracks and crevices on cliffs, as shown in the diagram below, where it can be reliably located and is easy to catch.

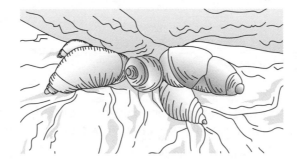

In an attempt to estimate how its population changes at one small cliff site, samples of snails were captured, marked and released on two separate occasions during a year of study, as shown in the table below.

Date of sampling	Number of snails captured and marked during sampling (M)	Number of snails in second sample (C)	Number of marked snails in second sample (R)	Population estimate (N)
March	1 435	1 725	195	–
September	1 400	–	250	7 000

(a) Use the population estimate formula N = MC/R to

    (i) calculate the population estimate for the March sample;   1

        Space for calculation

    _____

    (ii) calculate the number of second snails in the sample for September.   1

        Space for calculation

    _____

MARKS | DO NOT WRITE IN THIS MARGIN

8. (continued)

(b)   (i)   Describe an appropriate sampling method for a slow-moving mollusc with predictable behaviour at a small site such as this.

2

_____

_____

_____

(ii) Suggest an appropriate method for marking individual snails, and describe factors which should be considered when choosing the method to be used.

2

_____

_____

_____

MARKS | DO NOT WRITE IN THIS MARGIN

9. *Cyanea* is a genus of endemic flowering plants on the Hawaiian islands. *Cyanea* is thought to have co-evolved with species of Hawaiian honeycreepers and honeyeaters, which serve as pollinators of their flowers. The birds visit flowers to obtain nectar from nectaries within the bases of the flower tubes. As the birds probe the flowers with their beaks, pollen is brushed onto the feathering of their heads and can be carried to the next flower and rubbed off onto its stigmas. The diagram below shows the curvature and length of the flower tubes of two different *Cyanea* species and the heads of their main pollinators.

Cyanea superba

Drepanis pacifica

Cyanea fissa

Hemiatone sanguinea

(a) Explain what is meant by the term co-evolved. **1**

_____

_____

(b) Explain the advantage to *Drepanis pacifica* of its relationship with *Cyanea superba*. **2**

_____

_____

_____

(c) Describe how the Red Queen hypothesis can be used to explain the co-evolution of plants and their pollinators. **2**

_____

_____

_____

10. The dipper, *Cinclus cinclus*, is a small bird which lives by fast-flowing rocky streams in northern and western Britain. The birds feed on invertebrates which they pick from beneath the water surface. They build their nests on ledges or under rocky overhangs on steep river banks.

Dippers breed comparatively early in the year but the date of egg-laying varies. During 1987, observers in Wales monitored pairs of breeding dippers and recorded the date on which their first egg was laid. The pH of the water in the nest area was also measured. The graph below shows the results of their investigation.

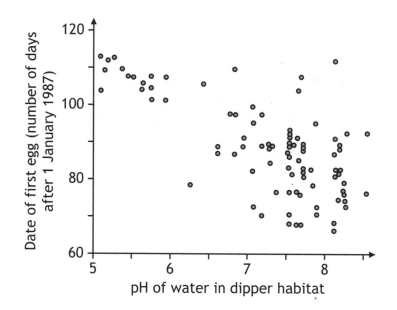

It was concluded that the less acidic the water in their habitat, the earlier in the year the birds' egg-laying date.

(a) The hypothesis being tested in the investigation above is that water pH affects egg-laying date in dippers.

Suggest **two** other factors which might be affecting egg-laying date in this species.

2

_____

_____

_____

MARKS | DO NOT WRITE IN THIS MARGIN

10. (continued)

(b) (i) The data can be attributed to correlation or causation.

From the data given, decide which you agree with and tick the appropriate box.

correlation ☐   causation ☐

Explain the reason for your choice.   **2**

_____

_____

_____

_____

_____

(ii) Compare the relationship between the pH of water or egg-laying date on pH 5.0–6.0 or pH 7.0–8.0.   **1**

_____

_____

_____

(c) Suggest **one** factor related to the validity and **one** factor related to the reliability of the experimental design which would have to be taken into account when making conclusions from the results.   **2**

Validity _____

_____

Reliability _____

_____

MARKS | DO NOT WRITE IN THIS MARGIN

11. Answer **either A or B** in the space below.

   A   Describe the primary, secondary, tertiary and quaternary levels of protein structure.   **8**

   OR

   B   Describe the role of photoreceptors in triggering a nervous impulse in animal eyes.   **8**

Space for answer

[END OF MODEL PAPER]

ADDITIONAL SPACE FOR ANSWERS AND ROUGH WORK

MARKS

DO NOT WRITE IN THIS MARGIN

ADDITIONAL SPACE FOR ANSWERS AND ROUGH WORK

ADVANCED HIGHER

2016

National
Qualifications
2016

X707/77/02

Biology
Section 1 — Questions

MONDAY, 9 MAY
9:00 AM – 11:30 AM

Instructions for the completion of Section 1 are given on *Page two* of your question and answer booklet X707/77/01.

Record your answers on the answer grid on *Page three* of your question and answer booklet.

Before leaving the examination room you must give your question and answer booklet to the Invigilator; if you do not, you may lose all the marks for this paper.

SECTION 1 — 25 marks

Attempt ALL questions

1. An experiment was set up to measure the activity of an enzyme using a substrate that produced a coloured product. The absorbance of the coloured product was measured using a colorimeter.

Which row in the table describes the variable being measured?

	discrete	continuous	qualitative	quantitative
A	✓		✓	
B	✓			✓
C		✓	✓	
D		✓		✓

2. The diagram below shows the cell types used in the production of monoclonal antibodies.

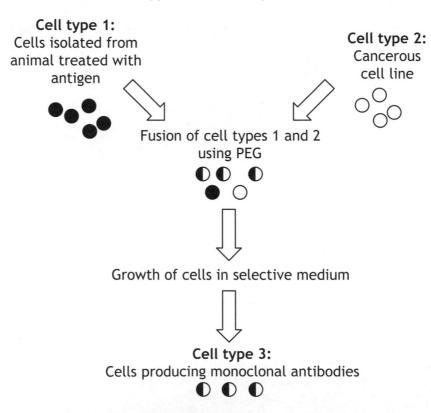

Cell type 1:
Cells isolated from animal treated with antigen

Cell type 2:
Cancerous cell line

Fusion of cell types 1 and 2 using PEG

Growth of cells in selective medium

Cell type 3:
Cells producing monoclonal antibodies

Which row in the table identifies cell types 1, 2 and 3?

	Cell type 1	Cell type 2	Cell type 3
A	B lymphocyte	myeloma	hybridoma
B	myeloma	hybridoma	B lymphocyte
C	hybridoma	myeloma	B lymphocyte
D	myeloma	B lymphocyte	hybridoma

3. The diagram below represents a transmembrane protein. Some of the amino acids in the protein have been identified.

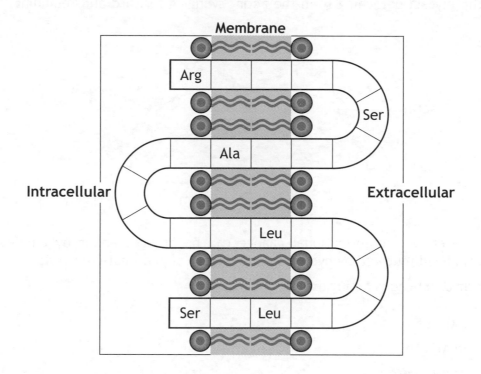

Which row in the table classifies the amino acids shown in this protein?

	Arginine (Arg)	Alanine (Ala)	Leucine (Leu)	Serine (Ser)
A	polar	hydrophobic	hydrophobic	polar
B	hydrophobic	polar	hydrophobic	polar
C	polar	hydrophobic	polar	hydrophobic
D	hydrophobic	polar	polar	hydrophobic

[Turn over

Questions 4 and 5 refer to the following information.

During muscle contraction, the protein myosin moves along an actin protein filament by the head of the myosin detaching from the actin, swinging forward and rebinding, as shown in the diagram.

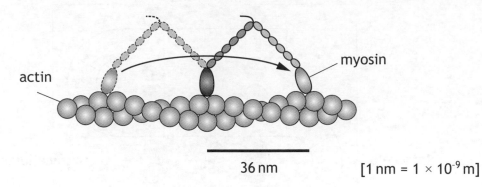

36 nm   [1 nm = 1 × 10^{-9} m]

4. This reversible conformational change can be brought about by binding of ATP to the myosin head followed by hydrolysis and release of phosphate and ADP.

   The myosin head is acting as

   A   a kinase

   B   an ATPase

   C   a proteinase

   D   a phosphatase.

5. When the myosin head detaches and swings forward it moves a distance of 36 nanometres (nm).

   Myosin has been observed to move at a speed of 18×10^3 nm s^{-1}.

   How many times will the myosin head detach and swing forward in one second?

   A   50

   B   200

   C   500

   D   2000

6. In animal rod cells rhodopsin absorbs a photon of light initiating the following cell events.

   1   nerve impulse is generated
   2   sufficient product formation is triggered
   3   activation of hundreds of G-protein molecules
   4   activation of hundreds of molecules of an enzyme

   The correct order of events is

   A   4, 2, 1, 3

   B   3, 4, 2, 1

   C   4, 3, 1, 2

   D   3, 2, 4, 1.

7. In multicellular organisms, only target cells respond to a specific signal because

    A   signalling molecules only come into contact with target cells

    B   only target cells have receptor molecules for the signalling molecule

    C   non-target cells do not respond when the signalling molecule binds to its receptor

    D   receptor molecules in non-target cells do not change conformation when the signalling molecule binds.

8. The hormone thyroxine is

    A   hydrophobic and unable to pass through the cell membrane

    B   hydrophilic and unable to pass through the cell membrane

    C   hydrophobic and able to pass through the cell membrane

    D   hydrophilic and able to pass through the cell membrane.

9. Biological molecules move over short distances by diffusion. The time taken for diffusion can be calculated using the equation below.

$$t = \frac{x^2}{2D}$$

    t = time taken (seconds)
    x = distance travelled by the diffusing molecule (cm)
    D = diffusion co-efficient (cm^2 per second)

    Acetylcholine is a neurotransmitter with a diffusion co-efficient of 4×10^{-6} cm^2 per second. The gap across the synapse is 5×10^{-6} cm wide.

    How many seconds would it take acetylcholine to cross the synapse?

    A   1·250

    B   6·250 × 10^{-6}

    C   3·125 × 10^{-6}

    D   1·600 × 10^{-6}

10. Type 1 diabetes is caused by

    A   excessive production of insulin

    B   loss of insulin receptor function

    C   failure of GLUT4 to respond to insulin binding

    D   insufficient production of insulin.

[Turn over

11. An enzyme-controlled reaction is taking place in optimum conditions in the presence of a large surplus of substrate.

Conditions can be altered by

1   increasing the temperature
2   adding a positive modulator
3   increasing enzyme concentration
4   increasing substrate concentration.

Product yield would be increased by

A   1 and 2

B   2 and 3

C   2 and 4

D   3 and 4.

12. At which phase of the cell cycle is the retinoblastoma protein phosphorylated allowing progression to the next phase of the cycle?

A   G1

B   S

C   G2

D   M

13. The diagram below shows possible outcomes for a cell following DNA damage. Protein X is involved in all three outcomes.

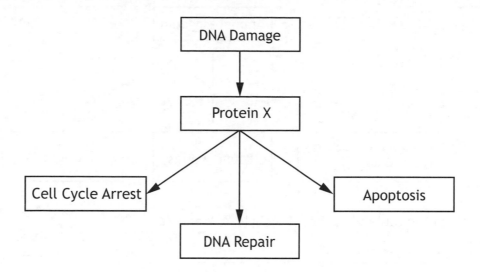

Protein X is

A   Rb

B   p53

C   Cdk

D   caspase.

14. Two reagents used in testing for the presence of carbohydrates are iodine solution, which turns blue-black in the presence of starch, and Benedict's solution, which turns brick red in the presence of maltose.

In an investigation of the breakdown of starch into maltose by the enzyme amylase, which of the following would be a positive control?

A   Maltose alone turns Benedict's solution brick red.

B   Starch treated with amylase turns Benedict's solution brick red.

C   Starch alone tested with Benedict's solution remains blue.

D   Starch treated with amylase does not change the colour of iodine solution.

[Turn over

15. The diagram below shows some phyla in the animal kingdom.

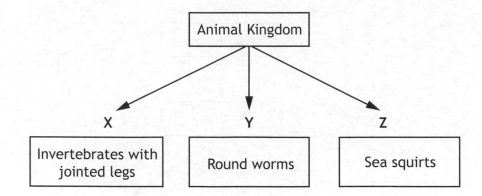

Which row in the table identifies the phyla X, Y and Z?

	Phylum		
	X	Y	Z
A	Chordata	Nematoda	Arthropoda
B	Arthropoda	Nematoda	Chordata
C	Nematoda	Arthropoda	Chordata
D	Arthropoda	Chordata	Nematoda

16. Which of the following descriptions of animal behaviour avoids the use of anthropomorphism?

A   In some primate species, alpha males often bully lower-ranking animals.

B   In late summer, worker bees like to visit heather flowers.

C   The grin on the chimpanzee's face showed that it was amused by the gesture.

D   The male moth is attracted to the female by the scent molecules that she emits.

17. A population of chafer beetles were damaging the tees and greens of a golf course. Results from a mark and recapture study suggested a population size that was too small to account for the extent of the damage caused.

One possible reason for this is that the

A   white paint used to mark the beetles washed off some of them before the recapture

B   white paint used to mark the beetles made them more visible to predators than unmarked beetles

C   total number of beetles in the recaptured sample was less than the number first captured and marked

D   marked beetles did not have enough time, after release, to spread out and mix with the rest of the population.

18. Ellis-van Creveld syndrome is a rare genetic condition. It is much more common in an isolated population in North America, which was founded by a small number of individuals, than in the general population.

The most likely explanation for this is

A   natural selection

B   sexual selection

C   random mutation

D   genetic drift.

19. The frequency of a given allele in a population is a measure of how common that allele is as a proportion of the total number of copies of all alleles at a specific locus. For a locus with one dominant allele (A) and one recessive allele (a), the frequency of the dominant allele (p) and the frequency of the recessive allele (q) can be used to calculate the genetic variation of a population using the equations below.

$$p + q = 1$$

p   = frequency of A allele
q   = frequency of a allele

$$p^2 + 2pq + q^2 = 1$$

p^2 = frequency of homozygous (AA) individuals
q^2 = frequency of homozygous (aa) individuals
2pq = frequency of heterozygous (Aa) individuals

If the allele frequency of the recessive allele is 0·7, the proportion of individuals that would be heterozygous is

A   0·09

B   0·21

C   0·42

D   0·49.

[Turn over

20. In the fruit fly *Drosophila melanogaster* the gene for eye colour is sex-linked. The allele for red eye (R) is dominant to the allele for white eye (r).

A cross between two flies produced the offspring shown in the table below.

Sex of offspring	Number with white eyes	Number with red eyes
female	23	22
male	21	22

The genotypes of the parents in this cross were

A $X^r X^r$ and $X^R Y$

B $X^R X^r$ and $X^r Y$

C $X^R X^r$ and $X^R Y$

D $X^R X^R$ and $X^r Y$.

21. Which row in the table best describes r-selected species?

	Number of offspring	Offspring survival rate	Parental care
A	many	low	little
B	few	high	extensive
C	many	high	extensive
D	few	low	little

22. Shags and cormorants both belong to the genus *Phalacrocorax*. They look very similar and nest near each other on the same cliffs. The table below shows the main components of each bird's diet.

Prey	Percentage composition of diet	
	Shag (Phalacrocorax aristotelis)	Cormorant (Phalacrocorax carbo)
sand eels	33	0
sprats	49	1
flatfish	1	26
shrimps	2	33
gobies	4	17
other fish	4	18

The data in the table show

A   competitive exclusion

B   competition within each species

C   resource partitioning

D   the fundamental niche of each species.

[Turn over

23. A species of parasitic wasp (*Nasonia vitripennis*) lays its eggs in the larvae of flies where the eggs develop. This species displays a behaviour called *"superparasitism"* where, following the laying of eggs by one wasp, a second wasp may superparasitise the same host by also laying its eggs.

Researchers investigated the effects of superparasitism on the brood size and sex ratio of offspring in this species. Results were compared to a control that had been parasitised only once. Researchers were able to distinguish between the offspring of the first and second wasp.

Results are shown in the table below.

Offspring	Degree of parasitism		
	Superparasitism		Single parasitism control
	Wasp 1	Wasp 2	
brood size	18 ± 3	17 ± 4	20 ± 2
percentage of males	7 ± 2	22 ± 4	6 ± 1

The following statements refer to the data in the table.

1 Superparasitism significantly increased the percentage of males produced by both wasp 1 and wasp 2.

2 Superparasitism significantly increased the percentage of males produced by wasp 2 only.

3 Superparasitism had no significant effect on brood size.

4 Superparasitism significantly decreased the brood size produced by wasps 1 and 2.

Which of these statements are valid conclusions supported by the data?

A   1 and 3

B   1 and 4

C   2 and 3

D   2 and 4

24. The statements below describe events that occur following the engulfing of a pathogen by a phagocyte of the mammalian immune system.

    P   long term survival of lymphocytes
    Q   antigen presentation to lymphocytes
    R   antibody production by lymphocytes
    S   clonal selection of B lymphocytes

The correct sequence in which these events occur is

A   Q, R, S, P

B   R, Q, P, S

C   S, Q, P, R

D   Q, S, R, P.

25. Florida scrubjays have evolved a co-operative breeding system in which helper birds assist breeding pairs in raising young. The table below compares the effect of helpers on the breeding success of birds that are either experienced or inexperienced breeders.

Breeding experience of breeding pairs	Average number of offspring reared	
	Without helpers	With helpers
inexperienced	1·24	2·20
experienced	1·80	2·38

Helpers increase the average number of offspring reared by inexperienced breeding pairs compared to experienced breeding pairs by

A   19%

B   23%

C   45%

D   60%.

[END OF SECTION 1. NOW ATTEMPT THE QUESTIONS IN SECTION 2
OF YOUR QUESTION AND ANSWER BOOKLET]

[BLANK PAGE]

DO NOT WRITE ON THIS PAGE

AH

National
Qualifications
2016

Mark

X707/77/01

**Biology
Section 1 — Answer Grid
and Section 2**

MONDAY, 9 MAY

9:00 AM – 11:30 AM

Fill in these boxes and read what is printed below.

Full name of centre

Town

Forename(s)

Surname

Number of seat

Date of birth

Day   Month   Year   Scottish candidate number

Total marks — 90

SECTION 1 — 25 marks

Attempt ALL questions.

Instructions for completion of Section 1 are given on *Page two*.

SECTION 2 — 65 marks

Attempt ALL questions.

A Supplementary Sheet for Question 1 is enclosed inside the front cover of this question paper. Write your answers clearly in the spaces provided in this booklet. Additional space for answers and rough work is provided at the end of this booklet. If you use this space you must clearly identify the question number you are attempting. Any rough work must be written in this booklet. You should score through your rough work when you have written your final copy.

Use **blue** or **black** ink.

Before leaving the examination room you must give this booklet to the Invigilator; if you do not you may lose all the marks for this paper.

SECTION 1 — 25 marks

The questions for Section 1 are contained in the question paper X707/77/02.

Read these and record your answers on the answer grid on *Page three* opposite.

Use **blue** or **black** ink. Do NOT use gel pens or pencil.

1.   The answer to each question is **either** A, B, C or D. Decide what your answer is, then fill in the appropriate bubble (see sample question below).

2.   There is **only one correct** answer to each question.

3.   Any rough working should be done on the additional space for answers and rough work at the end of this booklet.

Sample Question

The thigh bone is called the

      A   humerus

      B   femur

      C   tibia

      D   fibula.

The correct answer is **B** — femur. The answer **B** bubble has been clearly filled in (see below).

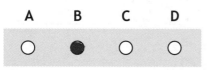

Changing an answer

If you decide to change your answer, cancel your first answer by putting a cross through it (see below) and fill in the answer you want. The answer below has been changed to **D**.

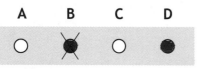

If you then decide to change back to an answer you have already scored out, put a tick (✓) to the **right** of the answer you want, as shown below:

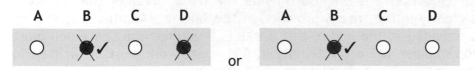

SECTION 1 — Answer Grid

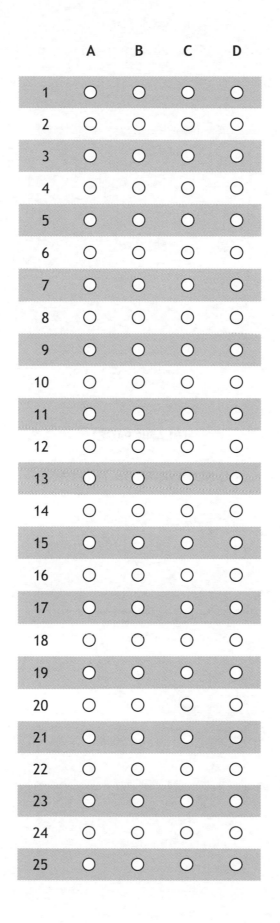

[Turn over

[BLANK PAGE]

DO NOT WRITE ON THIS PAGE

[Turn over for next question

DO NOT WRITE ON THIS PAGE

MARKS | DO NOT WRITE IN THIS MARGIN

SECTION 2 — 65 marks

Attempt ALL questions

It should be noted that question 11 contains a choice.

1. Read through the Supplementary Sheet for Question 1 before attempting this question.

   (a) **Refer to Figure 2 in the Supplementary Sheet for Question 1.**

      (i) Use the data to describe the egg-laying of uninfected mosquitoes.   **2**

      _____

      _____

      _____

      (ii) If the box plots were perfectly symmetrical, mean values for egg-laying would be very close to median values.

      State what can be deduced about the **mean** number of eggs laid by infected mosquitoes in relation to the median value.   **1**

      _____

      (iii) Describe the effect that *Plasmodium* infection has on the fecundity of mosquitoes used in the study.   **1**

      _____

   (b) **Refer to Figure 3 in the Supplementary Sheet for Question 1.**

      (i) The data shows that infection by *Plasmodium* appears to increase the longevity of female mosquitoes.

      Explain why the difference between the two groups can be regarded as significant.   **1**

      _____

      (ii) Suggest a benefit to the parasite of its vector living longer.   **1**

      _____

      _____

MARKS | DO NOT WRITE IN THIS MARGIN

1. **(continued)**

(c) **Refer to Figure 4 in the Supplementary Sheet for Question 1.**

(i) Explain what the lines of best fit indicate about the relationship between longevity and fecundity in both infected and uninfected mosquitoes.

2

_____

_____

_____

_____

(ii) State, with justification, whether or not this data is reliable.

1

_____

_____

_____

[Turn over

MARKS | DO NOT WRITE IN THIS MARGIN

2. Scientists have reported that neurons produced in cell culture from human stem cells have the potential to function when grafted into the site of a spinal injury in rats.

(a) State why the cell culture medium in which the neurons were cultured should contain serum.

1

_____

(b) Scientists used a haemocytometer to perform a cell count to calculate the number of stem cells that developed into neurons.

The diagram below represents a sample from a culture placed in a haemocytometer and viewed under a microscope.

The grid is **0·1 mm** in depth.

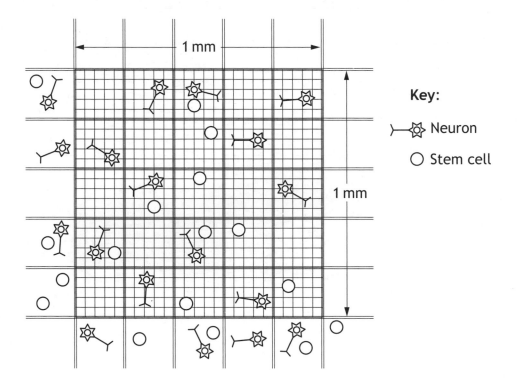

(i) Calculate the number of **neurons** in 1 cm³ of the culture.

Space for calculation

1

_____ neurons

(ii) Suggest **one** disadvantage of cell counts performed using the haemocytometer.

1

_____

MARKS | DO NOT WRITE IN THIS MARGIN

2. **(continued)**

(c) Bright field microscopy was used to view the cells grafted into the site of spinal injury.

State another type of biological material that can be viewed using bright field microscopy.   **1**

_____

(d) In studies involving animals, state **one** way in which harm to the animals can be minimised.   **1**

_____

[Turn over

MARKS DO NOT WRITE IN THIS MARGIN

3. Multiple sclerosis (MS) is a neurological condition in which the body's immune system destroys the myelin sheath that surrounds and insulates nerve axons.

A clinical study was carried out into the effects of a new drug *interferon beta-1b* for this condition. A randomised trial, with a negative control group (placebo), was carried out across four different health centres. During the study patients were given one of three treatments: 0·00 mg (placebo), 0·05 mg or 0·25 mg interferon. The patients administered the drug themselves at home.

The study measured how effective the drug was by asking patients to record any worsening of symptoms after 2 years of treatment. The study involved 372 patients aged 18-50 years. Fourteen patients dropped out before completing the trial.

Patients' results are shown in Table 1.

Table 1

Level of interferon beta-1b in treatment (mg)	Proportion of patients reporting no worsening of symptoms after 2 years of treatment (%)
0·00	16
0·05	18
0·25	25

At one health centre 52 patients were MRI scanned every 6 weeks to monitor any new damage to nerve tissue. The results are shown in Table 2.

Table 2

Level of interferon beta-1b in treatment (mg)	Proportion of patients showing new nerve damage (%)
0·00	29
0·05	no data recorded
0·25	6

(a) Identify the independent variable in this trial.

1

_____

(b) This trial was carried out *in vivo*.

State **one** advantage of this type of trial.

1

_____

_____

MARKS | DO NOT WRITE IN THIS MARGIN

3. **(continued)**

(c) Explain why a placebo group was included in this trial.   **1**

_____

_____

(d) Suggest **one** way in which the results of the trial may not be reliable.   **1**

_____

_____

(e) Describe an ethical issue that the researchers would need to consider before this trial.   **1**

_____

_____

(f) Suggest **two** conclusions that can be drawn from the results of this trial.   **2**

Conclusion 1 _____

_____

Conclusion 2 _____

_____

[Turn over

4. Sickle cell anaemia is an inherited blood disorder that reduces the ability of red blood cells to transport oxygen round the body by changing the structure of haemoglobin.

In sickle cell anaemia, the primary structure of a haemoglobin subunit is altered; the amino acid glutamic acid is substituted by the amino acid valine.

The structures of glutamic acid and valine are shown below.

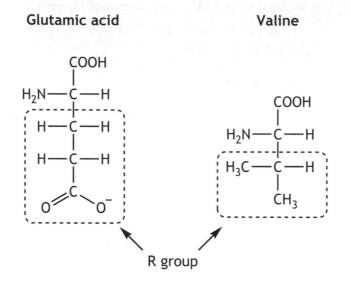

(a) State the class of amino acids to which valine belongs.   1

_____

(b) Identify **one** type of secondary structure shown in the haemoglobin molecule in the figure below.   1

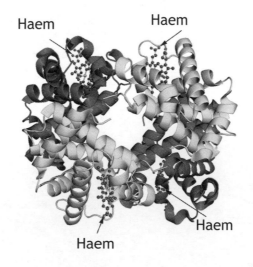

_____

MARKS | DO NOT WRITE IN THIS MARGIN

4. **(continued)**

(c) Explain the term cooperativity in relation to oxygen binding to haemoglobin.

1

_____

_____

(d) The graph below shows the oxygen saturation of haemoglobin at different oxygen pressures for an individual with normal haemoglobin and for another individual with sickle cell haemoglobin.

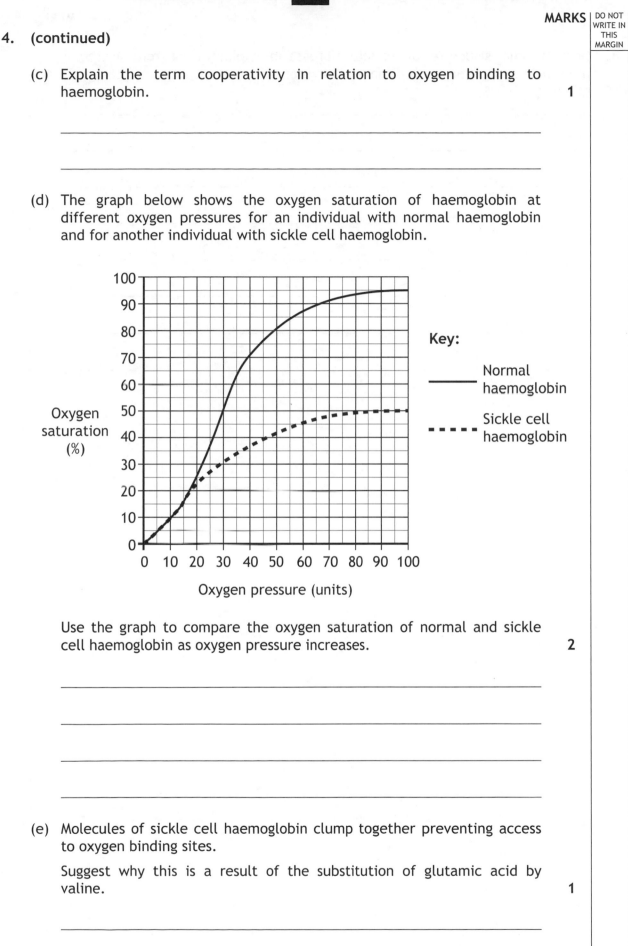

Use the graph to compare the oxygen saturation of normal and sickle cell haemoglobin as oxygen pressure increases.

2

_____

_____

_____

_____

(e) Molecules of sickle cell haemoglobin clump together preventing access to oxygen binding sites.

Suggest why this is a result of the substitution of glutamic acid by valine.

1

_____

_____

[Turn over

MARKS | DO NOT WRITE IN THIS MARGIN

5. Describe the structure of spindle fibres and explain their role in the movement of chromosomes during cell division.

4

_____

_____

_____

_____

_____

_____

_____

_____

MARKS | DO NOT WRITE IN THIS MARGIN

6. The sodium potassium pump (Na/KATPase) is a membrane protein found in animal cells.

(a) Give **one** function of sodium potassium pumps.

_____

_____

1

(b) Describe the role of ATP in altering the affinity of the pump for sodium ions (Na^+).

_____

_____

_____

2

(c) Digoxin is a chemical that inhibits the sodium potassium pump by binding to the potassium ion (K^+) binding site as shown in the diagram below.

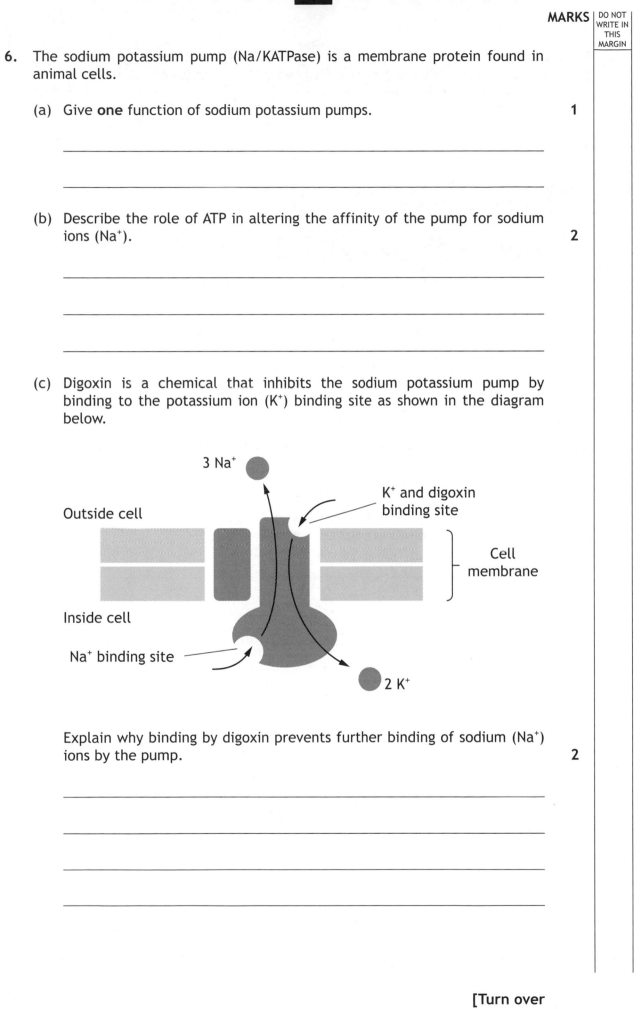

Explain why binding by digoxin prevents further binding of sodium (Na^+) ions by the pump.

_____

_____

_____

_____

2

[Turn over

MARKS | DO NOT WRITE IN THIS MARGIN

7. Binding of antidiuretic hormone (ADH) to its receptor on the plasma membrane of kidney collecting duct cells triggers the recruitment of water channel proteins as shown below.

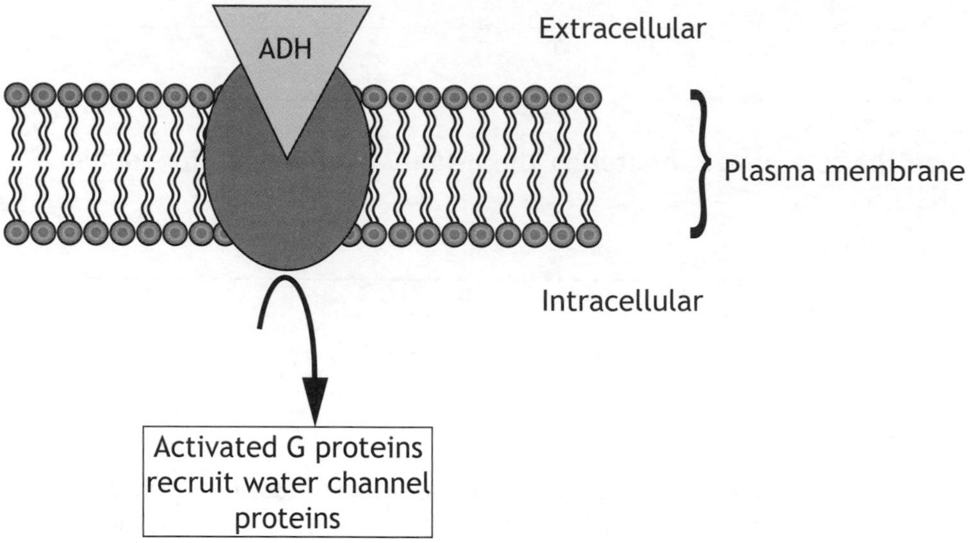

(a) (i) Name the water channel protein involved in this process.   1

_____

(ii) Name the process by which a response within the cell is triggered by the binding of ADH to its cell surface receptor.   1

_____

MARKS | DO NOT WRITE IN THIS MARGIN

7. **(continued)**

(b) A urine output of greater than 0·05 litres per kg body mass per day is considered diagnostic of diabetes insipidus. The bar chart below shows the urine output over 6 days of a 70 kg individual being investigated for diabetes insipidus. During days 3 and 4 the individual was treated with the drug *desmopressin*, a synthetic form of ADH.

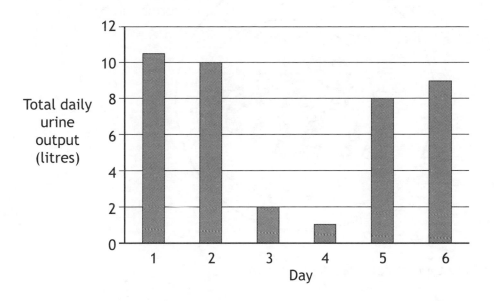

(i) Use the data to confirm that a diagnosis of diabetes insipidus is correct for this individual.

Space for calculation

1

_____

_____

_____

(ii) Give evidence from the graph that supports the conclusion that *desmopressin* is an effective treatment.

1

_____

_____

(iii) Diabetes insipidus results from failure to recruit water channel proteins to the cell membrane.

Identify the cause of recruitment failure in this individual.

1

_____

[**Turn over**

MARKS | DO NOT WRITE IN THIS MARGIN

8. The diagram below shows the pairing of homologous chromosomes in a cell undergoing meiosis.

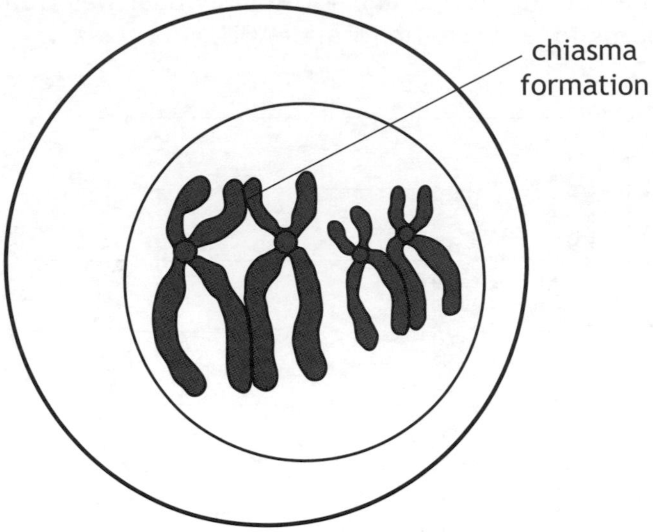

chiasma formation

(a) Name the type of cell that undergoes meiosis.   1

_____

(b) (i) Explain how the chiasma formation between the paired homologous chromosomes shown in the diagram leads to variation.   2

_____

_____

_____

_____

     (ii) Name the process that ensures haploid gametes produced by meiosis contain a mixture of chromosomes of maternal and paternal origin.   1

_____

[Turn over for next question

DO NOT WRITE ON THIS PAGE

9. In 1971, biologists moved five adult pairs of Italian wall lizards (*Podarcis sicula*) from their small home island of Kopiste to the neighbouring small island of Mrcaru, which did not have a lizard population. On their return in 2005 Mrcaru was found to have a large population of *P. sicula* (confirmed by genetic analysis) with significantly larger heads and a greater bite force than the lizards from Kopiste. Their digestive systems were also found to contain microorganisms that assist in the breakdown of plant cell walls.

The summer diets of the two lizard populations are shown below.

Key:

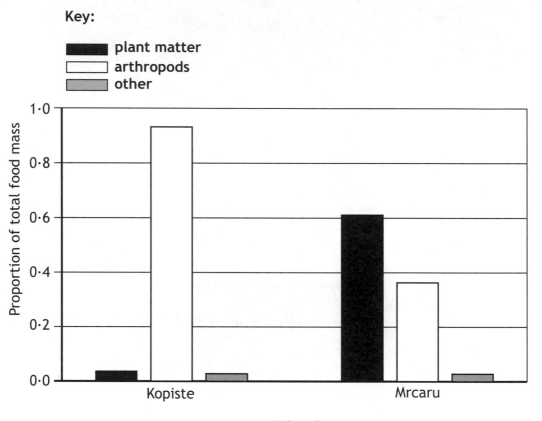

(a) Describe the most significant change in the summer diet of the lizards on Mrcaru.   1

_____

_____

MARKS | DO NOT WRITE IN THIS MARGIN

9. (continued)

(b) (i) Explain how the information supports the conclusion that the changes to the lizard population on Mrcaru were the result of natural selection.

2

_____

_____

_____

_____

(ii) Evolution of the lizards on Mrcaru occurred very rapidly.

State **one** factor that can increase the rate of evolution.

1

_____

(c) This study involved taking representative samples of the lizard populations on the two islands.

State **one** feature of a representative sample.

1

_____

_____

[Turn over

MARKS | DO NOT WRITE IN THIS MARGIN

10. The Figures below show male and female capercaillies (*Tetrao urogallus*) which are found in some Scottish pine forests. Males are much larger and darker than females and the breast feathers of the male have a metallic green sheen.

male capercaillie

female capercaillie

(a) State the term used to indicate the different body forms of males and females belonging to this species.   **1**

_____

(b) Capercaillies are a lekking species. Males perform displays during which they fan their tails, hold their wings down and make a variety of sounds. These features, which are attractive to females, are thought to serve as honest signals.

   (i) Explain what is meant by a lekking species.   **1**

_____

_____

   (ii) Explain why this display is often given as an example of sexual selection.   **1**

_____

_____

   (iii) If the display provides honest signals, state the benefit that may be obtained by females receiving these signals.   **1**

_____

_____

MARKS | DO NOT WRITE IN THIS MARGIN

10.　(continued)

(c)　Peacocks are the males of another lekking bird species, *Pavo cristatus*, whose natural habitat is the dense forests of South-East Asia. As well as the visual stimulus of a tail-feather display, peacocks, during mating, can emit a distinctive "hoot". These hoots are loud enough to be heard by other females, out of sight of the lek, who may be attracted by the calls and provide the dominant males at the lek with additional mating partners.

　(i)　Suggest why auditory stimuli are advantageous to species inhabiting forest ecosystems.

1

_____

_____

　(ii)　Recent research has found that some peacocks emit hoots in the complete absence of females at the lek. Females are still attracted to the lek by these sounds. Such "solo" hoots have been described as "dishonest signals".

Explain what is meant by a "dishonest signal" in this behaviour.

1

_____

_____

[Turn over for next question

MARKS | DO NOT WRITE IN THIS MARGIN

11. Answer **either A or B** in the space below and on *Page twenty-five*.

 A Discuss reproduction under the following headings:

   (i) costs and benefits of sexual reproduction; 4

   (ii) asexual reproduction as a successful reproductive strategy. 5

OR

 B Discuss endoparasitic infections under the following headings:

   (i) difficulties involved in their treatment and control; 7

   (ii) benefits of improved parasite control to human populations. 2

Labelled diagrams may be used where appropriate.

MARKS | DO NOT WRITE IN THIS MARGIN

SPACE FOR ANSWER FOR QUESTION 11

[END OF QUESTION PAPER]

ADDITIONAL SPACE FOR ANSWERS AND ROUGH WORK

ADDITIONAL SPACE FOR ANSWERS AND ROUGH WORK

[BLANK PAGE]

DO NOT WRITE ON THIS PAGE

National
Qualifications
2016

X707/77/11

Biology
Supplementary Sheet

MONDAY, 9 MAY

9:00 AM – 11:30 AM

Supplementary Sheet for Question 1

1. Malaria is caused by unicellular parasites in the genus *Plasmodium*. Figure 1 shows the life cycle of the parasite with respect to its human and mosquito hosts.

   Figure 1

   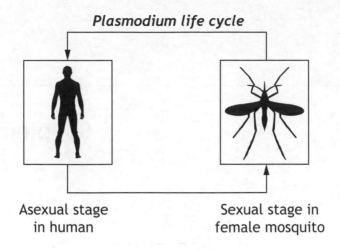

   Plasmodium life cycle

   Asexual stage   Sexual stage in
   in human   female mosquito

   Malaria is a well-researched tropical disease of humans, but less is known about the effects of the parasite on its mosquito vector.

   The parasite *Plasmodium relictum* causes malaria in birds. A recent study has been carried out to investigate the effects of this parasite on the mosquito *Culex pipiens*. In particular, two aspects were investigated: fecundity (number of eggs laid) and longevity (measured as survival after egg laying) of the mosquitoes.

   In Figure 2, box-and-whisker plots show the total egg production by large numbers of uninfected and infected female mosquitoes.

   Figure 2

   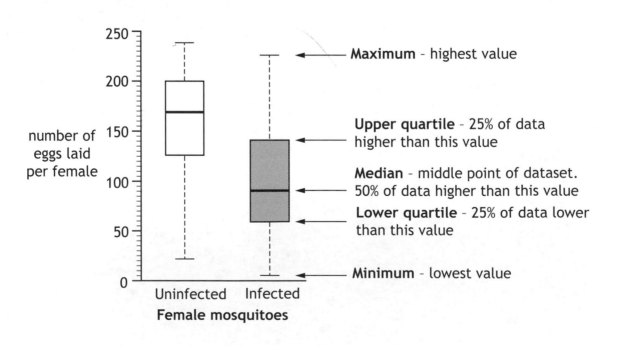

1.   (continued)

Figure 3 shows mean survival times after egg laying for uninfected and infected female mosquitoes.

Figure 3

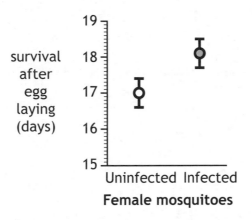

Fecundity and longevity were measured in the same individual female mosquitoes to see if there was a relationship between the two variables.

The lines of best fit for mosquito survival against the number of eggs each female laid were plotted for uninfected females and infected females.

This data is shown in Figure 4.

Figure 4

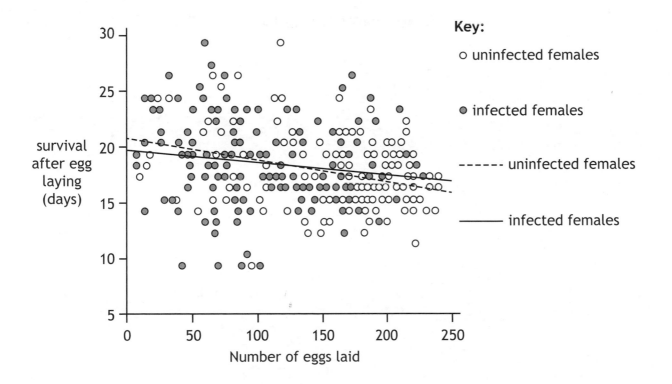

[BLANK PAGE]

DO NOT WRITE ON THIS PAGE

ADVANCED HIGHER

Answers

ADVANCED HIGHER BIOLOGY
2015 SPECIMEN QUESTION PAPER

Section 1

Question	Response	Mark
1.	A	1
2.	C	1
3.	B	1
4.	A	1
5.	D	1
6.	B	1
7.	C	1
8.	A	1
9.	C	1
10.	A	1
11.	C	1
12.	D	1
13.	C	1
14.	D	1
15.	A	1
16.	B	1
17.	C	1
18.	C	1
19.	B	1
20.	D	1
21.	D	1
22.	B	1
23.	A	1
24.	D	1
25.	C	1

Section 2

Question			Expected response	Max mark	Additional guidance
1.	(a)		Proteome	1	
	(b)		• Drosha not working • miRNA/precursor not processed/cut • No (micro)RNA strand for RISC OR RISC can't bind (m)RNA • (RNA) interference reduced/translation is left on	2	Any two
	(c)	(i)	Cell growth/cell increases in mass	1	
		(ii)	62·5	1	
		(iii)	More KO cells in G1 and fewer in S (and G2 + M)　(1) Differences are significant (only) in G1 and S/error bars don't overlap in G1 and S　(1) OR If comparing only G1 bars or only S bars, then must point out significant difference　(1)	2	Comparison can be made via data but data must be correct
	(d)	(i)	(After induction of differentiation) • in KO cells it is (generally) lower than normal cells • it increases in normal cells (over time) • in KO cells + one from below 　— no trend 　— decreases from day 8 　— increases (to day 8) then decreases (It = expression = level of marker)	1	Any one
		(ii)	• In normal cells, as differentiation increases self-renewal decreases **OR** converse (must link the two graphs/processes) • In KO/abnormal cells, **both** processes decrease after day 8 • in KO/abnormal cells, **both** processes increase to day 8 • in KO/abnormal cells, self-renewal remains higher and differentiation remains lower **than normal**	2	Any two
2.	(a)		• GABA is a ligand/substance that can bind to protein • The channel is a protein that opens in response to GABA/ligand binding • Chloride passes through the protein when GABA is bound	2	Any two
	(b)		Transduction	1	
	(c)	(i)	Chloride movement is (generally) greater at any GABA conc. if drug present	1	
		(ii)	Changes the conformation of the GABA site	1	
		(iii)	(Make the cell more negative inside so) increase the membrane potential	1	

Question			Expected response	Max mark	Additional guidance
3.	(a)		Concentration of ATP solution	1	
	(b)	(i)	Freshness of meat/whether meat has been frozen/ temperature of storage/incubation time/time before measurement/thickness of strip	1	Any one
		(ii)	• Storage of meat may cause damage to muscle proteins so muscle contraction would be less with less fresh meat • Freezing meat may damage muscle fibres so less contraction would be measured • Freezing meat may preserve muscle proteins so more contraction would be measured • As storage temperature increases protein damage may increase so less muscle contraction would be measured • Increasing incubation time/time before measurement will give more time for ATP to diffuse so more muscle contraction would be measured • ATP will diffuse more slowly through thicker strips which could mean the solution does not reach all the fibres so less contraction will be measured	1	Any one Explanation of effect must match with chosen confounding variable
	(c)		Not reliable (1) No independent replication/whole experiment was only carried out once OR Only one measurement for each chop at each concentration of ATP (1)	2	
	(d)		Negative control	1	
	(e)		May have prevented a representative sample being selected	1	
4.	(a)		The training will have no effect on GLUT 4 content of muscle	1	
	(b)		ND UT is baseline GLUT 4 **and** training does not produce significant increase (1) D UT is (significantly) lower GLUT 4 than baseline and exercise generates significant increase (1)	2	
	(c)		Type 1 diabetes is failure to produce insulin whereas type 2 diabetes is loss of insulin receptor function/ failure to respond to insulin	1	
5.	(a)		Hydrophobic	1	
	(b)		Thyroxine receptor protein is blocking transcription/ thyroxine binding removes repression of genes (1) More NaKATPase in membrane so more energy expenditure/higher metabolic rate (1)	2	
	(c)	(i)	People need to be treated for several weeks before metabolic rate reaches normal	1	
		(ii)	Starting metabolic rate is different for each individual	1	

Question			Expected response	Max mark	Additional guidance
6.	(a)		Retinal	1	
	(b)		• Excited rhodopsin activates G protein which in turn activates many enzyme molecules • Enzyme molecules cause closure of ion channels/ catalyse the removal of molecules that keep channels open • Inward leakage of positive ions/Na$^+$ and Ca$^+$ is halted so membrane potential increases • Hyperpolarisation/increasing charge stimulates nerve impulse	2	Any two
	(c)		Wide range of wavelengths absorbed/high degree of amplification from single photon	1	
7.	(a)		Men have one allele of the haemophilia gene whereas women have two alleles (of the haemophilia gene) **(1)** Recessive allele causing haemophilia not masked in men **(1)**	2	
	(b)		Daughter 100% **(1)** Son 50% **(1)**	2	
	(c)	(i)	Prevents a double dose of gene products (coded by the X chromosome) that might disrupt cellular function	1	
		(ii)	Inactivation of X chromosomes is random so this patient must have (by chance) more cells that have inactivated the unaffected allele/fewer cells that inactivated the affected allele	1	
8.			• homologous chromosomes pair (during meiosis I) • breakage and re-joining of DNA strands • at chiasmata • shuffles sections of DNA between homologous chromosomes • allows the recombination of alleles • (as) linked genes are separated	4	Any four
9.	(a)	(i)	Does not itself actively transmit parasite to another species	1	
		(ii)	Waterborne dispersal stage	1	
	(b)		• Mimic host antigens to evade detection • Modify host immune response to reduce chances of destruction • Antigenic variation allows rapid evolution to overcome host immune cell clonal selection	1	Any one
	(c)		Co-evolution of related species that interact frequently/ closely **(1)** Change in traits of one species acts as a selection pressure on the other species **(1)**	2	

Question			Expected response	Max mark	Additional guidance
10.	(a)		300%	1	
	(b)	(i)	Reduction in abundance of named species due to increase in seal population/physical damage/trampling **(1)** Increase in abundance of *Prasiola crispa* due to reduced competition for space/greater tolerance of trampling **(1)**	2	
		(ii)	Loss of plants gives areas of bare rock OR Not all plant species counted	1	
	(c)	(i)	Carried out in a way that minimises impact on environment OR Consideration of rare/vulnerable species	1	
		(ii)	Population being sampled is split into sub-populations	1	
11.	A		1. Immune surveillance by white blood cells 2. T lymphocytes recognise antigens from pathogen 3. Antigens (from pathogen) displayed on the surface of infected cells 4. T lymphocytes destroy infected cells 5. T lymphocytes induce apoptosis 6. Phagocytes present antigens to lymphocytes 7. B lymphocytes produce specific antibodies 8. T lymphocytes/B lymphocytes amplified by clonal selection 9. A different lymphocyte is produced/selected for each antigen 10. Long-term survival of some members of T lymphocyte/B lymphocyte clones 11. Surviving lymphocytes act as immunological memory cells	8	Any eight
	B		1. Female choice assesses male fitness 2. Females assess honest signals (to assess fitness) 3. Fitness explained in terms of advantageous genes/low parasite burdens 4. Display behaviour of lekking species 5. Successful strategies of dominant and satellite males (in lekking species) 6. Example of lekking behaviour described 7. Male—male rivalry: large size/weaponry 8. Increases access to females through conflict 9. Behaviour of sneaker males 10. Importance of sign stimuli and fixed action pattern in birds/fish 11. Example of sign stimuli and fixed action pattern described	8	Any eight

ADVANCED HIGHER BIOLOGY MODEL PAPER

Section 1

Question	Response	Mark
1.	B	1
2.	A	1
3.	A	1
4.	B	1
5.	C	1
6.	B	1
7.	A	1
8.	A	1
9.	C	1
10.	B	1
11.	D	1
12.	C	1
13.	D	1
14.	B	1
15.	A	1
16.	B	1
17.	B	1
18.	C	1
19.	C	1
20.	D	1
21.	A	1
22.	D	1
23.	A	1
24.	D	1
25.	D	1

Section 2

Question			Expected response	Max mark
1.	(a)		Glucose transport is greater in control red muscle cells than in control white muscle cells (1) When treated with insulin, glucose transport is increased more in red muscle cells than in white muscle cells (1)	2
	(b)		The standard error bars for red muscle and white muscle overlap showing that differences in average values are not significant	1
	(c)	(i)	High level of GLUT1 in plasma membrane and none in internal membranes before treatment	1
		(ii)	No effect on GLUT1 in plasma membrane **OR** Slight decrease in GLUT1 in plasma membrane	1
		(iii)	Decreases GLUT4 in plasma membrane and increases it in internal membranes in (both muscle types)	1
	(d)		Figure 3 shows that GLUT4 increases in the plasma membrane and decreases in the internal membranes in response to insulin (1) This change occurs in both muscle types but is greater in red than white muscle (1)	2
	(e)		In type 2 diabetes, insulin receptors are not so sensitive (1) So less GLUT4 is recruited into the plasma membrane (1)	2

Question			Expected response	Max mark
2.	(a)	(i)	Photon drives electron flow **(1)** And the energy from the electrons drives the H^+ flow **(1)**	2
		(ii)	Diffusion (through protein X) down the concentration gradient	1
	(b)		ATP synthase **(1)** Allows the plant to trap energy in a chemical form/as ATP **(1)**	2
3.	(a)		Oxygen/carbon dioxide/ others	1
	(b)	(i)	Water **(1)** Move from a high concentration to a lower concentration **(1)**	2
		(ii)	Voltage-gated channels respond to changes in ion concentration	1
		(iii)	Phosphorylation provides energy for active transport (of sodium/potassium)	1
4.	(a)		Expose/immunise them to the appropriate antigen	1
	(b)		So that the hybrid makes antibodies specific to the appropriate antigen **AND** Produces an immortal cell line	1
	(c)		Prevention **OR** Diagnosis **OR** Treatment of disease [Any 2 for 1 mark each]	2
5.	(a)		Can kill infected cells	1
	(b)	(i)	DNAase **(1)** Proteinase/caspases **(1)**	2
		(ii)	Increased Bcl-2 so less inhibition of apoptosis **(1)** Tumour cells grow in the absence of apoptosis **(1)**	2
	(c)		DNA damage	1

Question			Expected response	Max mark
6.			1. Physical barriers such as skin/conjunctiva/ hair 2. Chemical secretions such as stomach acid/ mucus/tears/wax 3. Inflammatory responses 4. Phagocytosis/ phagocytes destroy(s) infected cells 5. Natural killer (NK) cells 6. Destroy abnormal cells [Any 4 for 1 mark each]	4
7.	(a)		Sexual dimorphism	1
	(b)	(i)	Occupy leks which are very attractive to females	1
		(ii)	Don't use energy defending territories but gain access to females that visit leks	1
	(c)	(i)	Mating with more than one partner	1
		(ii)	Makes her less easy to be seen by predators which raid the nests of ground-nesting birds	1
8.	(a)	(i)	12 694	1
		(ii)	1 250	1
	(b)	(i)	Systematic/methodical search **(1)** Over a set/defined time period **(1)**	2
		(ii)	That the paint should ultimately wear off/be non-indelible/ non-permanent **(1)** **OR** Choose different paint colours for different sampling times **(1)**	2
9.	(a)		Evolution in pairs of species which interact frequently/closely	1
	(b)		Reduction in competition for food/nectar **(1)** Only it has a beak that can reach the nectar within the *Cyanea superba* flowers **(1)**	2

Question			Expected response	Max mark
	(c)		Plant flower shape change provides selection pressure **(1)** Which acts on the pollinator to alter the direction of natural selection of beak length **(1)**	2
10.	(a)		Rainfall **OR** Temperature **OR** Invertebrate food availability (others) **[Any 2 for 1 mark each]**	2
	(b)	(i)	Correlation pH may be indirectly related to date of first egg **(1)** and may be caused by another variable such as rainfall/pollution **(1)** **OR** Causation pH is directly related to date of first egg **(1)** as shown by a line of best fit through the data points **(1)**	3
		(ii)	Between pH 5.0 and 6.0 there is a clear relationship between the pH and date of first egg **AND** between pH 7.0 and 8.0 no clear trend/relationship can be seen between pH and date of first egg	2

Question			Expected response	Max mark
	(c)		Validity — effect of monitoring/disturbance on dipper behaviour **OR** Bias in the selection of nest to study, such as the remoteness of the nest sites **(1)** Reliability — number of birds studied **OR** Seasons involved **OR** Areas of the country covered (others) **(1)**	2
11.	A		1. Primary structure is the sequence of amino acids in the polypeptide 2. Secondary structure is the hydrogen bonding along the backbone 3–4. Which give alpha/α helices **OR** parallel/antiparallel beta/β sheets **OR** beta/β turns **[Any 2]** 5. Folding of polypeptide into a tertiary structure 6. Caused by bonding through interactions between R groups in hydrophobic regions 7. And/or ionic bonds 8. van der Waals interactions 9. Disulfide bridges 10. Prosthetic groups are non-protein groups 11. Which bind tightly to protein and are necessary for its function 12. Quaternary structure is formed in proteins with several sub-units **[Any 8 for 1 mark each]**	8

Question			Expected response	Max mark
B			1. Rhodopsin is the light-sensitive protein (in rod cells) 2. (Rhodopsin) is retinal combined with opsin 3. Cone cells are sensitive to specific/different wavelengths/colours 4. (In cone cells) different forms of opsin (combine with retinal) 5. Very high degree of amplification in rod cells 6. Results in sensitivity in low light intensities 7. Photon stimulates rhodopsin 8. A cascade of proteins amplifies the signal 9. Hundreds of G protein molecules are activated 10. This activates hundreds of molecules of enzyme 11. Enzymes generate product molecules 12. Sufficient product/threshold of product leads to a nerve impulse **[Any 8 for 1 mark each]**	8

ADVANCED HIGHER BIOLOGY 2016

Section 1

Question	Answer	Max mark
1.	D	1
2.	A	1
3.	A	1
4.	B	1
5.	C	1
6.	B	1
7.	B	1
8.	C	1
9.	C	1
10.	D	1
11.	B	1
12.	A	1
13.	B	1
14.	A	1
15.	B	1
16.	D	1
17.	D	1
18.	D	1
19.	C	1
20.	B	1
21.	A	1
22.	C	1
23.	C	1
24.	D	1
25.	C	1

Section 2

Question			Expected answer(s)	Max mark	Additional guidance
1.	(a)	(i)	Statement relating to quartiles e.g. 25% lay more than/UQ is 200 eggs e.g. 50% lay more than 170 eggs **OR** Median value is 170 eggs (laid) e.g. 75% lay more than/LQ is 125 eggs e.g. 50% lay between 125 and 200 OR Equivalents 'in opposite direction' OR Range of eggs (laid) is between 20 and 240/range of eggs (laid) is 220 OR Minimum and maximum values are 20 and 240 eggs (laid) OR No. of eggs (laid) is very variable **(Any two)**	2	• If no reference to *'eggs'* deduct one mark only **Not:** • average = median • 'average median'
		(ii)	Mean number of eggs laid/it is higher (than the median) OR Mean is greater than 90	1	
		(iii)	(Infection) reduces (fecundity)	1	Fecundity = no. of eggs laid
	(b)	(i)	Error bars do not overlap	1	
		(ii)	Increases (chance of)/more time for transmission (of parasite) OR More time for (parasite) reproduction	1	Transmission = spread/passed on to host **Not:** • allows transmission • reference to humans **Ignore:** • reference to intermediate/ definitive host

Question			Expected answer(s)	Max mark	Additional guidance
	(c)	(i)	Negative correlation between survival and the number of eggs laid OR Mosquitoes that lay smaller numbers of eggs live longer **(1)** Relationship is more negatively correlated in uninfected mosquitoes OR As fecundity increases the decrease in longevity is greater in uninfected mosquitoes **(1)**	2	Accept converse. Accept converse. Accept converse.
		(ii)	not reliable because many points lie far from the line OR reliable because a large sample was used	1	
2.	(a)		(Serum) provides <u>growth factor(s)</u>	1	**Negates:** • nutrients
	(b)	(i)	110,000	1	
		(ii)	Dead cells are not distinguished from live cells (unless stained) OR Small cells difficult to locate OR Numbers obtained are only an estimate OR Time-consuming OR Clumping of cells **(Any one)**	1	
	(c)		(Thin sections of) tissue OR Whole/unicellular organism OR Parts of organism **(Any one)**	1	**Not:** • named example of organism • named examples of parts of organisms
	(d)		**Replacement** (with another biological system, eg cell culture) OR **Reduction** (in no. used) OR **Refinement** (re techniques) **(Any one)**	1	Any description/ example should relate to one of the concepts.

Question			Expected answer(s)	Max mark	Additional guidance
3.	(a)		Level/quantity of interferon (beta-1b)	1	interferon beta-1b = drug
	(b)		Allows (overall) effect of drug on (whole) organism/body to be observed OR Allows (possible) side effects to be seen OR Shows effects on non-target cells OR Nerve cells difficult to grow *in vitro* (Any one)	1	**Not:** • Reference to ecological validity
	(c)		Provides results in the absence of the drug OR Gives baseline against which effect of drug can be measured/compared OR Allows comparison between drug and absence of drug OR Shows drug was responsible for effect OR Allows measurement of psychological effect (of treatment) (Any one)	1	**Not:** • Presence of drug = treatment
	(d)		Patients may not (remember to) take drug OR Patients may not inject/administer drug correctly/effeactively OR May be different numbers in the three groups OR Some patients pulled out (before completing trial) OR Not all patients were (MRI) scanned/no scan data for 0·05 mg OR Patient self-assessment (is subjective/may be recorded incorrectly) OR Small sample size (Any one)	1	

Question			Expected answer(s)	Max mark	Additional guidance
	(e)		Informed consent OR Permission from patient to use results/data OR Right to withdraw OR Confidentiality OR Justification of research OR (Consider possible) risk/harm/side effects to patient **(Any one)**	1	
	(f)		Drug prevents/reduces **worsening** of MS/symptoms OR Higher levels of drug more effective OR Drug reduces **new** nerve damage **(Any two)**	2	Interferon beta-1b = drug **Not only** reference to single data point for conclusions based on Table 1. **FOR Table 2: Accept** reference to single data point but **NOT** dose related trend.
4.	(a)		Hydrophobic/Non-polar	1	
	(b)		Alpha-helix Turn	1	
	(c)		Binding (to one subunit of one oxygen) makes the binding of other oxygen more likely	1	Correct reference to affinity change for binding/release of oxygen. **Not:** • binding to other Hbs
	(d)		At low pressures (below 15–20) there is no difference OR Comparison of (maximum) O_2 saturation at 90/100 pressure units (95 vs 50%) (1) At high pressures (15–20 upwards) increase for normal is greater than for sickle cell (1)	2	

Question			Expected answer(s)	Max mark	Additional guidance
	(e)		Valine has no charge (on R group) so (haemoglobin) molecules don't repel (one another) OR Hydrophobic interactions occur between (R groups of) valines (causing clumping) OR Glutamic acid has a charge (on R group) so (haemoglobin) molecules repel (Any one)	1	Interaction between non polar R groups of valines is equivalent to hydrophobic interactions between valines.
5.			1. Cell division requires remodelling of cytoskeleton 2. Spindle fibres made of microtubules 3. Composed of tubulin 4. (Composed of) hollow/ straight rods/cylinders/tubes **Maximum 2 marks from 1 to 4** 5. Attach to chromosomes/chromatids/centromeres/ kinetochores 6. Radiate from centrosome/microtubule organising centre/MTOC 7. Spindle fibres contract/shorten 8. Separate chromatids/(homologous) chromosomes **Maximum 2 marks from 5 to 8**	4	**Pt 6** Radiate = extend = grow = originate = made **Allow** Radiate from centriole but **NOT** Grow from/made by centriole.
6.	(a)		Maintaining osmotic balance OR Generation of ion (concentration) gradient AND one from: • For glucose symport (in small intestine) • In kidney tubules • For maintenance of resting potential (in cells/ neurons)	1	**Not:** • Maintain osmotic gradient
	(b)		Phosphorylation/conformational change (of pump) (1) **Lowers** affinity (for Na^+ ions) (1)	2	Conformational change must relate to ATP binding/ phosphorylation.
	(c)		Prevents binding of K^+ ions (1) Preventing de-phosphorylation OR Prevents (reversal of) conformational change OR Affinity for Na^+ ions (remains) low (1)	2	

Question			Expected answer(s)	Max mark	Additional guidance
7.	(a)	(i)	Aquaporin-(2)/AQP(2)	1	
		(ii)	Signal transduction	1	
	(b)	(i)	Urine output >3·5 litres (per day) on days without treatment/on day 1/2/5/6 OR Correct calculation of urine output per kg body mass on day 1/2/5/6 AND Stating value >0·05 litres/kg/day	1	Units required. e.g. from: day 1: 10·5/70 = ~0·15 day 2: 10/70 = 0·14 day 5: 8/70 = ~0·11 day 6: 9/70 = ~0·13
		(ii)	Urine production was <3·5 litres/<0·05 litres per kg on day 3/4 OR Urine production was <3·5 litres per day/<0·05 litres per kg per day during treatment OR Correct calculation showing urine output reduced to < critical level on day 3/4 (Any one)	1	'on day 3/4' **equivalent to** 'during treatment'
		(iii)	Failure to produce/lack of ADH	1	
8.	(a)		Gamete mother cell	1	
	(b)	(i)	Crossing over (at chiasmata) OR Breakage and rejoining of DNA/chromatids (at chiasmata) (1) (Leads to) exchange of DNA/alleles between (homologous) chromosomes OR New combinations of/recombination of alleles (of linked genes) (1)	2	
		(ii)	Independent assortment	1	
9.	(a)		(Much) greater proportion (of Mrcaru lizard's diet) is plant matter	1	
	(b)	(i)	(Mrcaru) lizards have micro-organisms to break down plant matter/greater bite force (1) These individuals AND are (better) **adapted** to new environment/eating plant matter/digesting plant matter OR have selective advantage/increased fitness (1)	2	suited ≠ adapted Accept description of *selective advantage*.

Question			Expected answer(s)	Max mark	Additional guidance
		(ii)	Short(er) generation time OR Warm(er) environment/climate/high(er) temperature OR High(er) selection pressure OR High(er) mutation rate OR Sexual reproduction/horizontal gene transfer **(Any one)**	1	
	(c)		Same mean as population as a whole OR Same degree of variation about/deviation from mean as the population (as a whole) **(Any one)**	1	
10.	(a)		Sexual dimorphism	1	
	(b)	(i)	Males gather/compete in (communal) area/lek (to display) AND Females assess/choose male OR To allow female choice	1	
		(ii)	(Display) increases **male's** chance of mating/passing on genes/reproducing OR (Display) increases **male's** breeding success	1	
		(iii)	(Surviving) offspring have increased fitness/more favourable characteristics OR High/greater number of surviving offspring	1	Characteristics = traits = genes = alleles
	(c)	(i)	(Sound) allows communication over (long) distance OR (Sound) overcomes difficulty of limited visibility OR (Sound) allows communication in spite of forest/trees limiting visual signals OR Allows female to locate male(s)/lek **(Any one)**	1	
		(ii)	(Dishonest as fake hoots) not indicating male mating success/fitness OR (Dishonest as fake hoots emitted when) females not present/no mating occurring	1	

Question			Expected answer(s)	Max mark	Additional guidance
11.	A	(i)	Costs/benefits of sexual reproduction: 1. Males/50% are unable to produce offspring OR Only females/50% able to produce offspring Only half of (each parent's) genome passed on (to offspring) 2. Disrupts successful (parental) genomes OR (Combinations of) beneficial alleles/traits lost 3. Increases (genetic) variation 4. (Variation) allows evolution/adaptation (in response to changing environment) 5. (Variation allows organism) to keep running in the Red Queen arms race (e.g. between parasite and host) **Maximum 4 marks from 1 to 6**	9	Produce offspring = reproduce Genome = genes = alleles = DNA = genetic information **Pt 2—NOT:** • traits
		(ii)	Asexual reproduction as a successful reproductive strategy: a. Successful genome passed on b. In narrow stable niches c. When recolonizing disturbed habitats d. Vegetative cloning in plants OR description of suitable example e. Parthenogenesis (in animals) OR description of example f. (Parthenogenesis) where parasite burden is low/climate is cool/parasite diversity is low g. (In organisms using asexual reproduction) horizontal gene transfer allows exchange of genetic material/increased variation h. Example of horizontal gene transfer **Maximum 5 marks from points a to h**		
	B	(i)	Difficulties involved in treatment and control: 1. Endoparasite defined as living within host 2. Rapid antigen change/high antigenic variation 3. Vaccines difficult to design/ produce 4. (Some) parasites difficult to culture (in vitro/laboratory) 5. Similarity between host and parasite **metabolism** 6. Difficult to find drugs only toxic to parasite 7. Difficulty associated with vector control OR Indirect transmission 8. Transmission rate high in tropical climate/overcrowded situations 9. Overcrowding (can occur) in refugee camps/rapidly growing cities (in LEDCs) 10. Difficult/expensive to improve sanitation **Maximum 7 marks from 1 to 10**	9	**Pt 8.** **Accept:** • Spread more rapidly • Overcrowding = high population density
		(ii)	Benefits of improved parasite control to human populations: a. Reduction in child mortality b. Improvements in child development/intelligence c. Body uses more resources for growth/development **Maximum 2 marks from a to c**		

Acknowledgements

Permission has been sought from all relevant copyright holders and Hodder Gibson is grateful for the use of the following:

Image © Artush/Shutterstock.com (SQP Section 2 page 19);
Image © petarg/Shutterstock.com (2016 Section 2 page 12);
Two images © Bildagentur Zoonar GmbH/Shutterstock.com (2016 Section 2 page 22).

Hodder Gibson would like to thank SQA for use of any past exam questions that may have been used in model papers, whether amended or in original form.